Student Solutions Manual and

College Physics

TENTH EDITION

VOLUME 1

Raymond A. Serway
Emeritus, James Madison University

Chris Vuille
Embry-Riddle Aeronautical University

Prepared by

John R. Gordon
Emeritus, James Madison University

Charles Teague
Emeritus, Eastern Kentucky University

Raymond A. Serway
Emeritus, James Madison University

Australia • Brazil • Mexico • Singapore • United Kingdom • United States

ISBN-13: 978-1-285-86625-3
ISBN-10: 1-285-86625-8

Cengage Learning
200 First Stamford Place, 4th Floor
Stamford, CT 06902
USA

Cengage Learning is a leading provider of customized learning solutions with office locations around the globe, including Singapore, the United Kingdom, Australia, Mexico, Brazil, and Japan. Locate your local office at: **www.cengage.com/global**.

Cengage Learning products are represented in Canada by Nelson Education, Ltd.

To learn more about Cengage Learning Solutions, visit **www.cengage.com**.

Purchase any of our products at your local college store or at our preferred online store **www.cengagebrain.com**.

Printed in the United States of America
1 2 3 4 5 6 7 17 16 15 14 13

PREFACE

This *Student Solutions Manual and Study Guide* has been written to accompany the textbook, *College Physics, Tenth Edition,* by Raymond A. Serway and Chris Vuille. The purpose of this ancillary is to provide students with a convenient review of the basic concepts and applications presented in the textbook, together with solutions to selected end-of-chapter problems. This is not an attempt to rewrite the textbook in a condensed fashion. Rather, emphasis is placed upon clarifying typical troublesome points and providing further practice in methods of problem solving.

Every textbook chapter has a matching chapter in this book and each chapter is divided into several parts. Very often, reference is made to specific equations or figures in the textbook. Each feature of this Study Guide has been included to insure that it serves as a useful supplement to the textbook. Most chapters contain the following components:

- **Notes from Selected Chapter Sections:** This is a summary of important concepts, newly defined physical quantities, and rules governing their behavior.

- **Equations and Concepts:** This is a review of the chapter, with emphasis on highlighting important concepts and describing important equations and formalisms.

- **Suggestions, Skills, and Strategies:** This section offers hints and strategies for solving typical problems that the student will often encounter in the course. In some sections, suggestions are made concerning mathematical skills that are necessary in the analysis of problems.

- **Review Checklist:** This is a list of topics and techniques the student should master after reading the chapter and working the assigned problems.

- **Solutions to Selected End-of-Chapter Problems:** Solutions are given for selected odd-numbered problems that were chosen to illustrate important concepts in each textbook chapter.

- **Tables:** A table of some Conversion Factors is provided on the inside front cover, and a list of selected Physical Constants is printed on the inside back cover.

An important note concerning significant figures: In preparing the problem solutions for this manual, we have attempted to carefully follow the rules regarding significant figures presented in Chapter 1 of the textbook. We sincerely hope that this *Student Solutions*

Manual and Study Guide will be useful to you in reviewing the material presented in the text and in improving your ability to solve problems and score well on exams. We welcome any comments or suggestions which could help improve the content of this study guide in future editions; and we wish you success in your study.

John R. Gordon
Harrisonburg, VA

Charles Teague
Richmond, KY

Raymond A. Serway
Leesburg, VA

Acknowledgments

We are indebted to everyone who contributed to this *Student Solutions Manual and Study Guide to Accompany College Physics, Tenth Edition*.

We wish to express our sincere appreciation to the staff of Cengage Learning who provided necessary resources and coordinated all phases of this project. Special thanks go to Chris Robinson (Product Assistant), Ed Dodd (Development Editor), and Charles Hartford (Senior Product Manager).

Our appreciation goes to our reviewer, Susan English, Durham Technical Community College. Her careful reading of the manuscript and checking the accuracy of the problem solutions contributed in an important way to the quality of the final product. Any errors remaining in the manual are the responsibility of the authors.

It is a pleasure to acknowledge the excellent work of the staff of MPS Limited, a Macmillan Company for assembling and typing this manual and preparing diagrams and page layouts. Their technical skills and attention to detail added much to the appearance and usefulness of this volume.

Finally, we express our appreciation to our families for their inspiration, patience, and encouragement.

Suggestions for Study

We have seen a lot of successful physics students. The question, "How should I study this subject?" has no single answer, but we offer some suggestions that may be useful to you.

1. Work to understand the basic concepts and principles before attempting to solve assigned problems. Carefully read the textbook before attending your lecture on that material. Jot down points that are not clear to you, take careful notes in class, and ask questions. Reduce memorization to a minimum. Memorizing sections of a text or derivations does not necessarily mean you understand the material.

2. After reading a chapter, you should be able to define any new quantities that were introduced and discuss the first principles that were used to derive fundamental equations. A review is provided in each chapter of the Study Guide for this purpose, and the marginal notes in the textbook (or the index) will help you locate these topics. You should be able to correctly associate with each physical quantity the symbol used to represent that quantity (including vector notation, if appropriate) and the SI unit in which the quantity is specified. Furthermore, you should be able to express each important principle or equation in a concise and accurate prose statement. Perhaps the best test of your understanding of the material will be your ability to answer questions and solve problems in the text or those given on exams.

3. Try to solve plenty of the problems at the end of the chapter. The worked examples in the text will serve as a basis for your study. This Study Guide contains detailed solutions to about twelve of the problems at the end of each chapter. You will be able to check the accuracy of your calculations for any odd-numbered problem, since the answers to these are given at the back of the text.

4. Besides what you might expect to learn about physics concepts, a very valuable skill you can take away from your physics course is the ability to solve complicated problems. The way physicists approach complex situations and break them down into manageable pieces is widely useful. Starting in Section 1.10, the textbook develops a general problem-solving strategy that guides you through the steps. To help you remember the steps of the strategy, they are called *Conceptualize, Categorize, Analyze,* and *Finalize*.

General Problem-Solving Strategy

Conceptualize

- The first thing to do when approaching a problem is to *think about* and *understand* the situation. Read the problem several times until you are confident you understand what is being asked. Study carefully any diagrams, graphs, tables, or photographs that accompany the problem. Imagine a movie, running in your mind, of what happens in the problem.

- If a diagram is not provided, you should almost always make a quick drawing of the situation. Indicate any known values, perhaps in a table or directly on your sketch.

- Now focus on what algebraic or numerical information is given in the problem. In the problem statement, look for key phrases such as "starts from at rest" ($v_i = 0$), "stops" ($v_f = 0$), or "freely falls" ($a_y = -g = -9.80 \text{ m/s}^2$). Key words can help simplify the problem.

- Next focus on the expected result of solving the problem. Exactly what is the question asking? Will the final result be numerical or algebraic? If it is numerical, what units will it have? If it is algebraic, what symbols will appear in it?

- Incorporate information from your own experiences and common sense. What should a reasonable answer look like? What should its order of magnitude be? You wouldn't expect to calculate the speed of an automobile to be 5×10^6 m/s.

Categorize

- Once you have a really good idea of what the problem is about, you need to *simplify* the problem. Remove the details that are not important to the solution. For example, you can often model a moving object as a particle. Key words should tell you whether you can ignore air resistance or friction between a sliding object and a surface.

- Once the problem is simplified, it is important to *categorize* the problem. How does it fit into a framework of ideas that you construct to understand the world? Is it a simple *plug-in problem,* such that numbers can be simply substituted into a definition? If so, the problem is likely to be finished when this substitution is done. If not, you face what we can call an *analysis problem*—the situation must be analyzed more deeply to reach a solution.

- If it is an analysis problem, it needs to be categorized further. Have you seen this type of problem before? Does it fall into the growing list of types of problems that you have solved previously? Being able to classify a problem can make it much easier to lay out a plan to solve it. For example, if your simplification shows that the problem can be treated as a particle moving under constant acceleration and you have already solved such a problem (such as the examples in Section 2.6), the solution to the new problem follows a similar pattern.

Analyze

- Now, you need to analyze the problem and strive for a mathematical solution. Because you have already categorized the problem, it should not be too difficult to select relevant equations that apply to the type of situation in the problem. For example, if your categorization shows that the problem involves a particle moving under constant acceleration, Equations 2.9 to 2.13 are relevant.

- Use algebra (and calculus, if necessary) to solve symbolically for the unknown variable in terms of what is given. Substitute in the appropriate numbers, calculate the result, and round it to the proper number of significant figures.

Finalize

- This final step is the most important part. Examine your numerical answer. Does it have the correct units? Does it meet your expectations from your conceptualization of the problem? What about the algebraic form of the result—before you substituted numerical values? Does it make sense? Try looking at the variables in it to see whether the answer would change in a physically meaningful way if they were drastically increased or decreased or even became zero. Looking at limiting cases to see whether they yield expected values is a very useful way to make sure that you are obtaining reasonable results.

- Think about how this problem compares with others you have done. How was it similar? In what critical ways did it differ? Why was this problem assigned? You should have learned something by doing it. Can you figure out what? Can you use your solution to expand, strengthen, or otherwise improve your framework of ideas? If it is a new category of problem, be sure you understand it so that you can use it as a model for solving future problems in the same category.

When solving complex problems, you may need to identify a series of sub-problems and apply the problem-solving strategy to each. For very simple problems, you probably don't need this whole strategy. But when you are looking at a problem and you don't know what to do next, remember the steps in the strategy and use them as a guide.

Work on problems in this Study Guide yourself and compare your solutions with ours. Your solution does not have to look just like the one presented here. A problem can sometimes be solved in different ways, starting from different principles. If you wonder about the validity of an alternative approach, ask your instructor.

5. We suggest that you use this Study Guide to review the material covered in the text and as a guide in preparing for exams. You can use the sections Chapter Review, Notes from Selected Chapter Sections, and Equations and Concepts to focus in on any points which require further study. The main purpose of this Study Guide is to improve upon the efficiency and effectiveness of your study hours and your overall understanding of physical concepts. However, it should not be regarded as a substitute for your textbook or for individual study and practice in problem solving.

TABLE OF CONTENTS

1

Introduction

NOTES ON SELECTED CHAPTER SECTIONS

1.1 Standards of Length, Mass, and Time

Systems of units commonly used are the **SI system**, in which the units of mass, length, and time are the kilogram (kg), meter (m), and second (s), respectively; the **cgs** or **gaussian system**, in which the units of mass, length, and time are the gram (g), centimeter (cm), and second, respectively; and the **U.S. customary system**, in which the units of mass, length, and time are the slug, foot (ft), and second, respectively.

The **meter** has been redefined several times. In October 1983, it was redefined to be the distance traveled by light in a vacuum during a time of 1/299 792 458 second.

The SI unit of mass, the **kilogram**, is defined as the mass of a specific platinum-iridium alloy cylinder kept at the International Bureau of Weights and Measures at Sèvres, France.

The **second** is now defined as 9 192 631 700 times the period of one oscillation of radiation from the Cesium-133 atom.

1.2 The Building Blocks of Matter

It is useful to view the atom as a miniature solar system with a dense, positively charged nucleus occupying the position of the Sun and negatively charged electrons orbiting like the planets. Occupying the nucleus are two basic entities, protons and neutrons. The **proton** is nature's fundamental carrier of positive charge; the **neutron** has no charge and a mass about equal to that of a proton. Even more elementary building blocks than protons and neutrons exist. Protons and neutrons are each now thought to consist of three particles called **quarks**.

1.3 Dimensional Analysis

Dimensional analysis makes use of the fact that dimensions can be treated as algebraic quantities. *Quantities can be added or subtracted only if they have the same dimensions.*

1.4 Uncertainty in Measurement and Significant Figures

A **significant figure** is a reliably known digit (other than a zero that is used to locate a decimal point).

When **multiplying several quantities**, the number of significant figures in the final result is the same as the number of significant figures in the least accurate of the quantities being multiplied, where "least accurate" means "having the lowest number of significant figures." The same rule applies to division.

When **numbers are added (or subtracted)**, the number of decimal places in the result should equal the smallest number of decimal places of any term in the sum (or difference).

Most of the numerical examples and end-of-chapter problems in your textbook will yield answers having either two or three significant figures.

1.5 Conversion of Units

Sometimes it is necessary to **convert units** from one system to another. A list of conversion factors can be found on the inside front cover of the ***Student Solutions Manual and Study Guide***.

1.6 Estimates and Order-of-Magnitude Calculations

Often it is useful to **estimate an answer** to a problem in which little information is given. In such a case we refer to the **order of magnitude** of a quantity, by which we mean the power of ten that is closest to the actual value of the quantity. *Usually, when an order-of-magnitude calculation is made, the results are reliable to within a factor of 10.*

1.7 Coordinate Systems

A **coordinate system** used to specify locations in space consists of:

- A fixed reference point O, called the origin
- A set of specified axes, or directions, with an appropriate scale and label on each of the axes
- Instructions that tell us how to label a point in space relative to the origin and axes

Cartesian (rectangular) coordinates and polar coordinates are two commonly used coordinate systems.

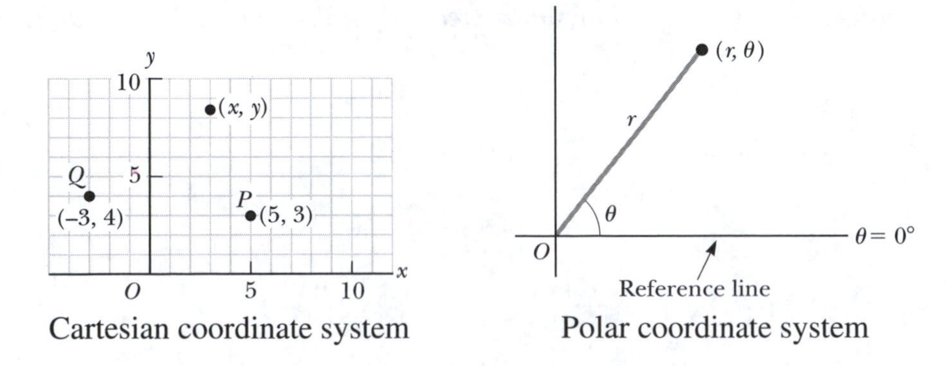

Cartesian coordinate system

Polar coordinate system

1.8 Trigonometry

You should review the basic trigonometric functions stated by Equations (1.1) and (1.2) in the **Equations and Concepts** section of this chapter.

1.9 Problem-Solving Strategy

In developing problem-solving strategies, the following steps will be helpful:

1. Read the problem carefully at least twice. Be sure you understand the nature of the problem before proceeding further.
2. Draw a suitable diagram with appropriate labels and coordinate axes, if needed.
3. Imagine what happens in the problem.
4. Identify the basic physical principle (or principles) involved, listing the knowns and unknowns.
5. Select (or derive) equation(s) as necessary to find the unknown quantities in terms of the known quantities.
6. Solve the equations for the unknowns symbolically (algebraically without substituting numerical values).
7. Substitute the given values with the appropriate units to obtain numerical values with units for the unknowns.
8. Check your answer against the following questions: Do the units match? Is the answer reasonable? Is the plus or minus sign proper or meaningful?

EQUATIONS AND CONCEPTS

The **three basic trigonometric functions** of an acute angle in a right triangle are the sine, cosine, and tangent.

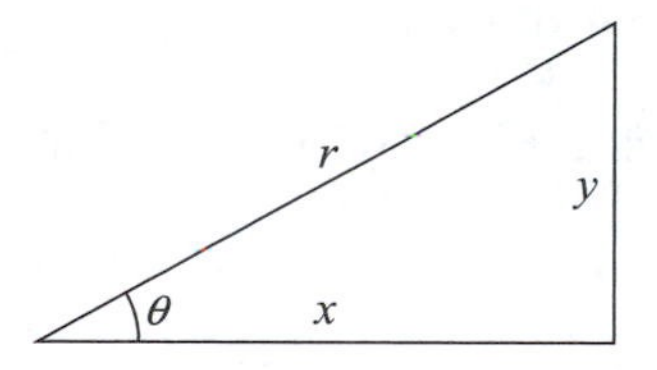

$$\sin\theta = \frac{\text{side opposite to }\theta}{\text{hypotenuse}} = \frac{y}{r}$$

$$\cos\theta = \frac{\text{side adjacent to }\theta}{\text{hypotenuse}} = \frac{x}{r} \qquad (1.1)$$

$$\tan\theta = \frac{\text{side opposite to }\theta}{\text{side adjacent to }\theta} = \frac{y}{x}$$

The **Pythagorean theorem** is an important relationship among the lengths of the sides of a right triangle.

$$r^2 = x^2 + y^2 \qquad (1.2)$$

SUGGESTIONS, SKILLS, AND STRATEGIES

Many mathematical symbols will be used throughout this book. Some important examples are:

$\propto$	denotes	a proportionality
$<$	means	"is less than"
$>$	means	"is greater than"
$<<$	means	"is much less than"
$>>$	means	"is much greater than"
$\cong$	indicates	approximate equality
$=$	indicates	equality
$\sim$	means	"is of the order of"
Δx ("delta x")	indicates	the change in a quantity x
$\lvert x \rvert$	means	the absolute value of x (always positive)
Σ (capital sigma)	represents	a sum. For example,

$$x_1 + x_2 + x_3 + x_4 + x_5 = \sum_{i=1}^{5} x_i$$

REVIEW CHECKLIST

- Discuss the units and standards of SI quantities: length, mass, and time.
- Derive SI units for physical quantities (e.g., force, velocity, volume, and acceleration) from units of the three basic quantities: length, mass, and time.
- Perform a dimensional analysis of an equation containing physical quantities whose individual units are known.
- Convert units from one system to another.
- Carry out order-of-magnitude calculations or "guesstimates."
- Describe the coordinates of a point in space using both Cartesian and polar coordinate systems.

SOLUTIONS TO SELECTED END-OF-CHAPTER PROBLEMS

5. Newton's law of universal gravitation is represented by

$$F = G\frac{Mm}{r^2}$$

where F is the gravitational force, M and m are masses, and r is a length. Force has the SI units $\text{kg}\cdot\text{m/s}^2$. What are the SI units of the proportionality constant G?

Solution

Solving Newton's law of universal gravitation for the proportionality constant, we find

$$G = \frac{F\cdot r^2}{Mm}$$

where F is a force having units of $[F]=[m]\cdot[\ell]/[t^2]=\text{M}\cdot\text{L/T}^2$, r is a length, while M and m are both masses. Therefore, the units of this proportionality constant are

$$[G]=\frac{[F]\cdot[r]^2}{[M]\cdot[m]}=\frac{(\cancel{\text{M}}\cdot\text{L/T}^2)\cdot\text{L}^2}{\cancel{\text{M}}\cdot\text{M}}=\frac{\text{L}^3}{\text{M}\cdot\text{T}^2}$$

In the SI system, the basic unit of length is the meter, the basic unit of mass is the kilogram, and the basic unit of time is the second. Thus, the SI units of this proportionality constant are

$$[G]=\frac{\text{m}^3}{\text{kg}\cdot\text{s}^2}$$ ◊

---

9. How many significant figures are there in (a) 78.9 ± 0.2, (b) 3.788×10^9, (c) 2.46×10^{-6}, (d) 0.003 2?

Solution

The significant figures in a measurement include all of the digits that are known reliably plus the first digit the experimenter had to estimate.

(a) The notation ±0.2 indicates an uncertainty of 2 units in the first decimal place. Thus, the number 78.9 contains 2 digits that are known reliably (the 7 and the 8) and one digit with some uncertainty (the 9). Therefore, it has three significant figures. ◊

(b) In scientific notation, the first part of the number (3.788 in this case) is always in the range 1 to 10 and contains all of the digits serving as significant figures. The second part of the number simply gives the power of 10 that is to multiply the first part. The number of significant figures contained in 3.788×10^9 is four. ◊

(c) Note that the first part of the number 2.46×10^{-6} contains three digits. Thus, as discussed in (b) above, the number of significant figures contained in 2.46×10^{-6} is three. ◊

(d) Expressing $0.003\,2$ in scientific notation gives 3.2×10^{-3}. (Note that in the original notation, the zeros merely serve to locate the decimal point.) Then, as discussed in part (b), observe that $0.003\,2$ contains two significant figures. ◊

---

13. The edges of a shoebox are measured to be 11.4 cm, 17.8 cm, and 29 cm. Determine the volume of the box retaining the proper number of significant figures in your answer.

Solution

The volume of a rectangular box is the product of its length, width, and height. Observe that in the given dimensions of this box, both 11.4 cm and 17.8 cm have three significant figures, while 29 cm contains only two significant figures. The rule of thumb given in the textbook concerning the significant figures in multiplication and division is as follows:

> In multiplying (dividing) two or more quantities, the number of significant figures in the product (quotient) is the same as the number of significant figures in the *least accurate* of the factors being combined, where *least accurate* means *having the lowest number of significant figures.*

The least accurate factor in the product of the three dimensions of the box is 29 cm with two significant figures. The volume of the box should therefore contain only two significant figures. This volume is then given by

$$V=\ell\cdot w\cdot h=(29\text{ cm})(17.8\text{ cm})(11.4\text{ cm})=5.9\times10^{3}\text{ cm}^{3}$$ ◊

---

17. A firkin is an old British unit of volume equal to 9 gallons. How many cubic meters are there in 6.00 firkins?

Solution

Note that the firkin is defined to be exactly 9 gallons, so we shall not consider the factor of 9 to limit the number of significant figures which may be retained in the final answer of this conversion calculation.

As in all conversion calculations, we start with the given data and multiply it by a series of ratios, with each ratio having a value of 1 (i.e., the numerator and denominator are equal to each other). We may multiply the original data by 1 as many times as desired, and while we may change its appearance, the final result is equal to the original value. Each ratio used is chosen to cancel some of the current units and move us one step closer to the desired units.

In the calculation given below, it is desired to convert cubic inches to cubic centimeters, and later to convert cubic centimeters to cubic meters. The table of conversion factors on the front flyleaf of your textbook does not include these conversion factors. Still, this table does tell us that $1\text{ in} = 2.54\text{ cm}$. Thus, $(1\text{ in})^2 = (2.54\text{ cm})^2$ and $(1\text{ in})^3 = (2.54\text{ cm})^3$. This means that we may use the ratio

$$\left(\frac{2.54\text{ cm}}{1\text{ in}}\right)^3 = \frac{(2.54\text{ cm})^3}{(1\text{ in})^3} = \frac{(2.54)^3\text{ cm}^3}{1\text{ in}^3}$$

in our calculation. Similarly, $1\text{ m} = 10^2\text{ cm}$, so

$$(1\text{ m})^2 = 1\text{ m}^2 = \left(10^2\text{ cm}\right)^2 = 10^4\text{ cm}^2$$

and

$$(1\text{ m})^3 = 1\text{ m}^3 = \left(10^2\text{ cm}\right)^3 = 10^6\text{ cm}^3$$

The calculation needed to convert 6.00 firkins to cubic meters is

$$6.00\text{ firkins} = 6.00\ \cancel{\text{firkins}}\left(\frac{9\ \cancel{\text{gal}}}{1\ \cancel{\text{firkin}}}\right)\left(\frac{231\ \cancel{\text{in}^3}}{1\ \cancel{\text{gal}}}\right)\left(\frac{(2.54)^3\ \cancel{\text{cm}^3}}{1\ \cancel{\text{in}^3}}\right)\left(\frac{1\text{ m}^3}{10^6\ \cancel{\text{cm}^3}}\right)$$

$$= 0.204\text{ m}^3$$ ◊

21. The diameter of a sphere is measured to be 5.36 in. Find (a) the radius of the sphere in centimeters, (b) the surface area of the sphere in square centimeters, and (c) the volume of the sphere in cubic centimeters.

Solution

(a) The radius of the sphere is one-half of its diameter. Hence, the radius is given by

$$r = \frac{d}{2} = \frac{5.36\text{ in}}{2} = 2.68\text{ in} = 2.68\ \cancel{\text{in}}\left(\frac{2.54\text{ cm}}{1\ \cancel{\text{in}}}\right) = 6.81\text{ cm}$$ ◊

(b) The surface area of the sphere is given by $A = 4\pi r^2$. Thus, we compute this area and convert it to units of square centimeters as

$$A = 4\pi r^2 = 4\pi(2.68\text{ in})^2 = 90.3\text{ in}^2\left(\frac{2.54\text{ cm}}{1\text{ in}}\right)^2 = 583\text{ cm}^2$$ ◊

(c) The volume of a sphere is given by $V = \frac{4}{3}\pi r^3$, so we compute the volume of this sphere and convert it to units of cubic centimeters as follows:

$$V = \frac{4}{3}\pi r^3 = \frac{4}{3}\pi(2.68\text{ in})^3 = 80.6\text{ in}^3\left(\frac{2.54\text{ cm}}{1\text{ in}}\right)^3 = 1.32\times 10^3$$ ◊

25. The amount of water in reservoirs is often measured in acre-ft. One acre-ft is a volume that covers an area of one acre to a depth of one foot. An acre is $43\ 560\ \text{ft}^2$. Find the volume in SI units of a reservoir containing 25.0 acre-ft of water.

Solution

To convert the volume of the reservoir from the given units of acre-ft to the standard units of volume in the SI system (m^3), it will be necessary to use multiple conversion factors. Each of these conversion factors will consist of a ratio which has a value of 1. The numerator and denominator of the ratio will be chosen to cancel some of the units we have and move us one step closer to the desired final units. Often, multiple conversion factors will be employed at the same time to do the entire conversion of units in a single step. We shall do the desired calculation in several steps, and then show how it could have been done in only one step.

$$V = 25.0\ \text{acre}\cdot\text{ft} = (25.0\ \cancel{\text{acre}}\cdot\text{ft})\left(\frac{43\ 560\ \text{ft}^2}{1\ \cancel{\text{acre}}}\right) = 1.09\times10^6\ \text{ft}^3$$

$$= 1.09\times10^6\ \text{ft}^3 = (1.09\times10^6\ \text{ft}^3)\left(\frac{1\ \text{m}}{3.281\ \text{ft}}\right) = 3.32\times10^5\ \text{ft}^2\cdot\text{m}$$

$$= 3.32\times10^5\ \text{ft}^2\cdot\text{m} = (3.32\times10^5\ \text{ft}^2\cdot\text{m})\left(\frac{1\ \text{m}}{3.281\ \text{ft}}\right) = 1.01\times10^5\ \text{ft}\cdot\text{m}^2$$

$$= 1.01\times10^5\ \text{ft}\cdot\text{m}^2 = (1.01\times10^5\ \text{ft}\cdot\text{m}^2)\left(\frac{1\ \text{m}}{3.281\ \text{ft}}\right) = 3.08\times10^4\ \text{m}^3 \quad \Diamond$$

To do this conversion in one step, it is helpful to realize that the ratio $1\ \text{m}/3.281\ \text{ft} = 1$ (since the numerator and the denominator are equal). Thus, $(1\ \text{m}/3.281\ \text{ft})^3 = 1^3 = 1$ Making use of this, we write

$$V = 25.0\ \text{acre}\cdot\text{ft} = (25.0\ \cancel{\text{acre}}\cdot\cancel{\text{ft}})\left(\frac{43\ 560\ \cancel{\text{ft}^2}}{1\ \cancel{\text{acre}}}\right)\left(\frac{1\ \text{m}}{3.281\ \cancel{\text{ft}}}\right)^3 = 3.08\times10^4\ \text{m}^3 \quad \Diamond$$

30. Estimate the number of people in the world who are suffering from the common cold on any given day. (Answers may vary. Remember that a person suffers from a cold for about a week.)

Solution

We assume that the average person will catch a serious cold about twice a year and is sick an average of seven days (1 week) each time. Thus, on average, each person suffers from a cold for two weeks out of each year (fifty two weeks).

The probability that a particular person will have a cold at any given time equals the percentage of the time that person is sick, or

$$\textit{probability of sickness} = \frac{2 \text{ weeks}}{52 \text{ weeks}} = \frac{1}{26}$$

With a current world population of approximately seven billion, the number of people we would expect to be suffering from a cold on any given day is

$$n = \textit{number sick} = (\textit{population})(\textit{probability of sickness}) = (7\times 10^9)\left(\frac{1}{26}\right) = 3\times 10^8$$

Our order of magnitude estimate is then $n \sim 10^8$. ◊

---

39. Two points are given in polar coordinates by $(r,\theta) = (2.00 \text{ m}, 50.0°)$ and $(r,\theta) = (5.00 \text{ m}, -50.0°)$, respectively. What is the distance between them?

Solution

Consider the sketch at the right. In this case, the first point has polar coordinates of

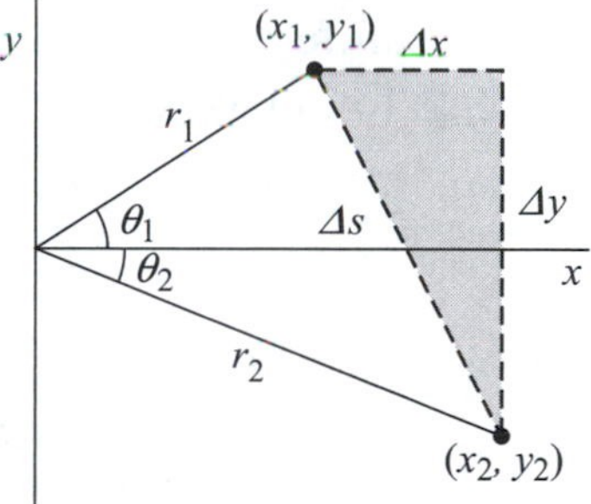

$$r_1 = 2.00 \text{ m}, \quad \theta_1 = 50.0°$$

and Cartesian coordinates of

$$x_1 = r_1 \cos\theta_1 = (2.00 \text{ m})\cos 50.0° = 1.29 \text{ m}$$

$$y_1 = r_1 \sin\theta_1 = (2.00 \text{ m})\sin 50.0° = 1.53 \text{ m}$$

Similarly, the polar and Cartesian coordinates for the second point are

$$r_2 = 5.00 \text{ m}, \quad \theta_2 = -50.0°$$

and

$$x_2 = r_2 \cos\theta_2 = (5.00 \text{ m})\cos(-50.0°) = 3.21 \text{ m}$$
$$y_2 = r_2 \sin\theta_2 = (5.00 \text{ m})\sin(-50.0°) = -3.83 \text{ m}$$

In moving from point 1 down to point 2, one must undergo a horizontal displacement of

$$\Delta x = x_2 - x_1 = 3.21 \text{ m} - 1.29 \text{ m} = +1.92 \text{ m}$$

and a vertical displacement of

$$\Delta y = y_2 - y_1 = (-3.83 \text{ m}) - 1.53 \text{ m} = -5.36 \text{ m}$$

Observe that these displacements are the legs of the shaded right triangle shown in the sketch. The straight line distance between the two points is the hypotenuse of this triangle and is given by the Pythagorean theorem as

$$\Delta s = \sqrt{(\Delta x)^2 + (\Delta y)^2} = \sqrt{(+1.92)^2 + (-5.36)^2} = 5.69 \text{ m} \qquad \Diamond$$

---

43. A high fountain of water is located at the center of a circular pool as shown in Figure P1.43. Not wishing to get his feet wet, a student walks around the pool and measures its circumference to be 15.0 m. Next, the student stands at the edge of the pool and uses a protractor to gauge the angle of elevation at the bottom of the fountain to be 55.0°. How high is the fountain?

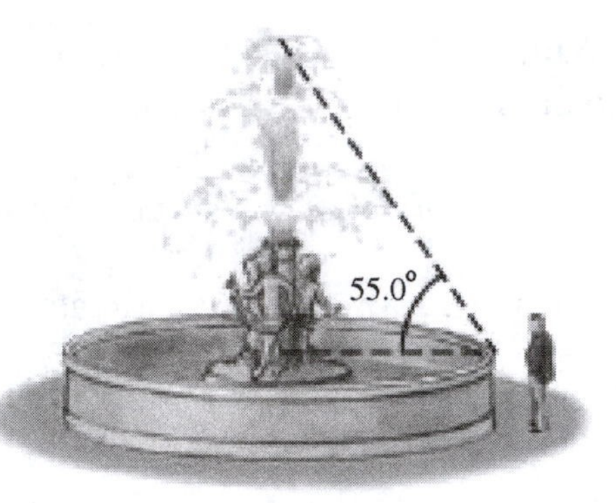

Figure P1.43

Solution

The circumference of the fountain is $C = 2\pi r$, so the radius is found to be

$$r = \frac{C}{2\pi} = \frac{15.0 \text{ m}}{2\pi} = 2.39 \text{ m}$$

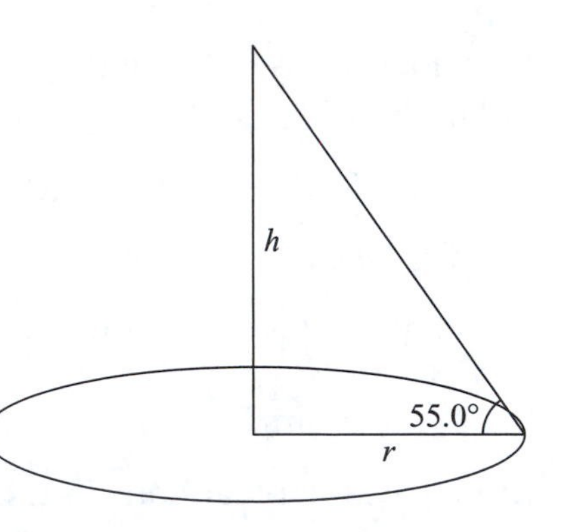

Now observe the sketch given at the right and note that

$$\tan(55.0°) = \frac{h}{r} = \frac{h}{2.39 \text{ m}}$$

Solving this expression for the height of the fountain gives

$$h = (2.39 \text{ m})\tan(55.0°) = 3.41 \text{ m} \qquad \Diamond$$

---

50. Refer to Problem 48. Suppose the mountain height is y, the woman's original distance from the mountain is x, and the angle of elevation she measures from the horizontal to the top of the mountain is θ. If she moves a distance d closer to the mountain and measures an angle of elevation ϕ, find a general equation for the height of the mountain y in terms of d, ϕ, and θ, neglecting the height of her eyes above the ground.

Solution

The drawing at the right summarizes the measurements described in the problem statement. The height of the mountain is y. The woman is initially distance x from the mountain, at which point she observes an angle of elevation θ. After walking distance d toward the mountain, she measures the new angle of elevation to be ϕ.

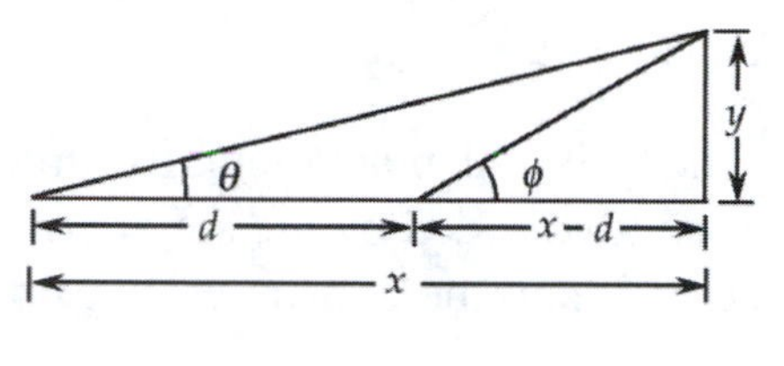

From the large triangle having base x and height y in this drawing, we observe that $\tan\theta = y/x$, or $x = y/\tan\theta$.

Looking again at the drawing, observing the small triangle having base $x-d$ and height y, we realize that $\tan\phi = y/(x-d)$. Solving this expression for the distance x gives $x = d + y/\tan\phi$.

We now have two different expressions for distance x. Equating them gives $y/\tan\theta = d + y/\tan\phi$, which can be solved for the height y as

$$y\left(\frac{1}{\tan\theta} - \frac{1}{\tan\phi}\right) = d$$

or

$$y\left(\frac{\tan\phi - \tan\theta}{\tan\theta\tan\phi}\right) = d$$

Thus, the desired expression for the height of the mountain is

$$y = \frac{d\cdot\tan\theta\cdot\tan\phi}{\tan\phi - \tan\theta}$$

◊

55. The displacement of an object moving under uniform acceleration is some function of time and the acceleration. Suppose we write this displacement as $s = ka^m t^n$, where k is a dimensionless constant. Show by dimensional analysis that this expression is satisfied if $m=1$ and $n=2$. Can this analysis give the value of k?

Solution

For the equation to be valid, we must choose values of m and n to make it dimensionally consistent.

Since s is a displacement, its dimensions are those of length, $[s]=\mathrm{L}$. The acceleration, a, is a length divided by the square of a time: $[a]=\mathrm{L}/\mathrm{T}^2$. The variable t has dimensions of time, $[t]=\mathrm{T}$, and the constant k has no dimensions.

Substituting these dimensions into the proposed equation yields

$$[s]=[a]^m[t]^n$$

$$\mathrm{L}=\left(\frac{\mathrm{L}}{\mathrm{T}^2}\right)^m \mathrm{T}^n=\mathrm{L}^m\,\mathrm{T}^{-2m}\,\mathrm{T}^n$$

or

$$\mathrm{L}^1\,\mathrm{T}^0=\mathrm{L}^m\,\mathrm{T}^{n-2m}$$

Note that the factor T^0 introduced on the left side of the equation is equal to 1, so its introduction does not change the equation. This equation can be true only if the powers of length L are the same on the two sides of the equation and, simultaneously, the powers of time T are the same on both sides. [*Note*: If the third basic unit, mass M, were present we would also require that its powers be identical on the two sides of the equation.]

Thus, we obtain a set of two simultaneous equations:

$$1=m \qquad \text{and} \qquad 0=n-2m$$

The solutions are seen to be

$$m=1 \qquad \text{and} \qquad n=2$$ ◊

This technique gives no information about the possible values of the dimensionless constant k. ◊

63. The nearest neutron star (a collapsed star made primarily of neutrons) is about 3.00×10^{18} m away from Earth. Given that the Milky Way galaxy (Fig. P1.63) is roughly a disk of diameter $\sim 10^{21}$ m and thickness $\sim 10^{19}$ m, estimate the number of neutron stars in the Milky Way to the nearest order of magnitude.

Solution

If we consider the Milky Way galaxy to be a disk with a diameter whose order of magnitude is $d \sim 10^{21}$ m and with a thickness on the order of $t \sim 10^{19}$ m, the order of magnitude of its volume must be

$$V_G = At = \left(\frac{\pi d^2}{4}\right)t \sim \frac{\pi}{4}\left(10^{21}\text{ m}\right)^2\left(10^{19}\text{ m}\right) = 7.9 \times 10^{60}\text{ m}^3 \quad \text{or} \quad V_G \sim 10^{61}\text{ m}^3$$

If, in the Milky Way galaxy, there is typically one neutron star in a spherical volume of radius $r = 3 \times 10^{18}$ m, the average galactic volume occupied by each neutron star is

$$V_1 = \frac{4}{3}\pi r^3 \sim \frac{4}{3}\pi\left(3 \times 10^{18}\text{ m}\right)^3 = 1 \times 10^{56}\text{ m}^3 \quad \text{or} \quad V_1 \sim 10^{56}\text{ m}^3$$

Our order of magnitude estimate for the number of neutron stars in the Milky Way galaxy is then

$$n = \frac{V_G}{V_1} \sim \frac{10^{61}\text{ m}^3}{10^{56}\text{ m}^3} \quad \text{or} \quad n \sim 10^5$$

◊

2

Motion in One Dimension

NOTES FROM SELECTED CHAPTER SECTIONS

2.1 Displacement

The displacement of an object, defined as its **change in position**, is given by the difference between its final and initial positions, or $x_f - x_i$. Displacement is an example of a vector quantity. *A vector is a physical quantity that requires a specification of both direction and magnitude.*

2.2 Velocity

Average speed is the ratio of the total distance traveled to the elapsed time of travel. *Speed is a scalar quantity and is always zero or positive.*

The **average velocity** of an object during the time interval t_i to t_f is equal to the slope of the straight line joining the initial and final points on a graph of the position of the object plotted versus time. *Velocity is a vector quantity and can be positive, negative, or zero.*

Instantaneous velocity (velocity at a specific time) is the slope of the line tangent to the position-time curve at a point P corresponding to the specified time.

The **instantaneous speed** of an object, which is a scalar quantity, is defined as the magnitude of the instantaneous velocity.

2.3 Acceleration

The **average acceleration** during a given time interval is defined as the change in velocity divided by the time interval during which this change occurs.

The **instantaneous acceleration** of an object at a certain time equals the slope of the tangent to the velocity vs. time graph at that instant of time.

2.4 Motion Diagrams

The **motion of an object** can be represented by a motion diagram with vectors indicating the direction and relative magnitude of the velocity and acceleration at successive time intervals. You should carefully study Active Figure 2.12 in the textbook. This figure illustrates the motion of a car in three different cases:

- Constant positive velocity, zero acceleration
- Positive velocity, positive acceleration
- Positive velocity, negative acceleration

2.5 One-Dimensional Motion with Constant Acceleration

This type of motion is important because it applies to many objects in nature. When an object moves with constant acceleration, the average acceleration equals the instantaneous acceleration. *Equations* (2.6) *through* (2.10) *may be used to solve any problem in one-dimensional motion with constant acceleration.*

2.6 Freely Falling Objects

A freely falling body is an object moving freely under the influence of gravity only, regardless of its initial motion.

It is important to emphasize that any freely falling object experiences an acceleration directed downward. *This is true regardless of the direction of motion of the object.* An object thrown upward or downward will experience the same acceleration as an object released from rest. *Once they are in free fall, all objects have an acceleration downward equal to the acceleration due to gravity.*

EQUATIONS AND CONCEPTS

The **displacement** Δx of an object is defined as its change in position. Equation (2.1) gives the displacement for a particle moving (in one-dimensional motion) from an initial position, x_i, to a final position, x_f. *For the case of one-dimensional motion, the usual vector notation is not required.*

$$\Delta x \equiv x_f - x_i \quad (2.1)$$

The **average velocity** of an object during a time interval is the ratio of its displacement to the time interval during which the displacement occurred.

$$\overline{v} \equiv \frac{\Delta x}{\Delta t} = \frac{x_f - x_i}{t_f - t_i} \quad (2.2)$$

The **instantaneous velocity** is defined as the limit of the average velocity as the time interval Δt goes to zero. *Note that the instantaneous velocity of an object might have different values from instant to instant.*

$$v \equiv \lim_{\Delta t \to 0} \frac{\Delta x}{\Delta t} \quad (2.3)$$

The **average acceleration** of an object during a time interval is the ratio of its change in velocity to the time interval during which the change in velocity occurs.

$$\bar{a} \equiv \frac{\Delta v}{\Delta t} = \frac{v_f - v_i}{t_f - t_i} \qquad (2.4)$$

The **instantaneous acceleration** is defined as the limit of the average acceleration as the time interval Δt goes to zero.

$$a \equiv \lim_{\Delta t \to 0} \frac{\Delta v}{\Delta t} \qquad (2.5)$$

The **equations of kinematics** can be used to describe one-dimensional motion with constant acceleration along the x-axis. Note that each equation shows a different relationship among physical quantities: initial velocity, final velocity, acceleration, time, and displacement. Also, in the forms that they are shown, it is assumed that $t_i = 0$, $t_f = t$, and for convenience, $\Delta x = x - x_0$. ***Remember, Equations 2.6–2.10 are only valid when the acceleration is constant.***

$$v = v_0 + at \qquad (2.6)$$

$$\bar{v} = \frac{v_0 + v}{2} \qquad (2.7)$$

$$\Delta x = \tfrac{1}{2}(v + v_0)t \qquad (2.8)$$

$$\Delta x = v_0 t + \tfrac{1}{2}at^2 \qquad (2.9)$$

$$v^2 = v_0^2 + 2a\Delta x \qquad (2.10)$$

The **motion of an object in free fall** (along the y-axis) can be described by Equations 2.6–2.10 when acceleration, a, is replaced by $-g$. *Note that since $a = -g$ in these four equations, the $+y$-axis is predefined to point upward.*

$$v = v_0 - gt$$

$$\Delta y = \tfrac{1}{2}(v + v_0)t$$

$$\Delta y = v_0 t - \tfrac{1}{2}gt^2$$

$$v^2 = v_0^2 - 2g\Delta y$$

The **maximum height** reached by an object thrown upward is determined by the initial height and the initial velocity.

$$y_{\max} = y_0 + \frac{v_0^2}{2g}$$

The **time to reach maximum height** is determined by the initial velocity.

$$t_{\max} = \frac{v_0}{g}$$

SUGGESTIONS, SKILLS, AND STRATEGIES

The following procedure is recommended for solving problems involving motion with constant acceleration.

1. Make sure all the units in the problem are consistent. That is, if distances are measured in meters, be sure that velocities have units of m/s and accelerations have units of m/s^2.

2. Choose a coordinate system and make a labeled diagram of the problem, including the directions of all displacements, velocities, and accelerations.

3. Make a list of all the quantities given in the problem and a separate list of those to be determined.

4. Select the equation(s) which will enable you to solve for the unknown(s).

5. Construct an appropriate motion diagram and check to see if your answers are consistent with the diagram of the problem.

REVIEW CHECKLIST

- Define the displacement and average velocity of a particle in motion. Define the instantaneous velocity and understand how this quantity differs from average velocity.
- Define average acceleration and instantaneous acceleration.
- Construct a graph of displacement versus time (e.g., for a function such as $x = 5 + 3t - 2t^2$) for a particle in motion along a straight line. From this graph, you should be able to determine both average velocity over a time period and the instantaneous velocity at a given time by calculating the slope of the tangent to the graph.
- Apply the kinematic equations of this chapter, Equations 2.6–2.10, to any situation where the motion occurs under constant acceleration.
- Describe what is meant by a body in free fall (one moving under the influence of gravity—where air resistance is neglected). Recognize that the equations of constant accelerated motion apply directly to a freely falling object and that the acceleration is then given by $a = -g$ (where $g = 9.80\ \text{m/s}^2$).

SOLUTIONS TO SELECTED END-OF-CHAPTER PROBLEMS

3. A person travels by car from one city to another with different constant speeds between pairs of cities. She drives for 30.0 min at 80.0 km/h, 12.0 min at 100 km/h, and 45.0 min at 40.0 km/h and spends 15.0 min eating lunch and buying gas. (a) Determine the average speed for the trip. (b) Determine the distance between the initial and final cities along the route.

Solution

(a) The distance traveled during each interval of the trip is given by $\Delta x = v(\Delta t)$, where v is the constant speed maintained during that interval and Δt is the duration of the interval. For the different intervals of the trip, we find

$$\Delta x_1 = (80.0\ \text{km/h})(30.0\ \text{min})(1\ \text{h}/60.0\ \text{min}) = 40.0\ \text{km}$$

$$\Delta x_2 = (100\ \text{km/h})(12.0\ \text{min})(1\ \text{h}/60.0\ \text{min}) = 20.0\ \text{km}$$

$$\Delta x_3 = (40.0\ \text{km/h})(45.0\ \text{min})(1\ \text{h}/60.0\ \text{min}) = 30.0\ \text{km}$$

$$\Delta x_4 = (0\ \text{km/h})(30.0\ \text{min})(1\ \text{h}/60.0\ \text{min}) = 0\ \text{km}$$

The total distance traveled during the trip is

$$\Delta x_{\text{total}} = \Delta x_1 + \Delta x_2 + \Delta x_3 + \Delta x_4 = 40.0 \text{ km} + 20.0 \text{ km} + 30.0 \text{ km} + 0 = 90.0 \text{ km}$$

and the total elapsed time is

$$\Delta t_{\text{total}} = \Delta t_1 + \Delta t_2 + \Delta t_3 + \Delta t_4 = (30.0 + 12.0 + 45.0 + 15.0) \text{ min} = 102 \text{ min}$$

Hence, the average speed for the trip is

$$\bar{v} = \frac{\Delta x_{\text{total}}}{\Delta t_{\text{total}}} = \frac{90.0 \text{ km}}{102 \text{ min}}\left(\frac{60.0 \text{ min}}{1 \text{ h}}\right) = 52.9 \text{ km/h} \qquad \Diamond$$

(b) As computed above, the total distance traveled between the initial and final cities of the trip is $\Delta x_{\text{total}} = 90.0$ km.

---

11. The cheetah can reach a top speed of 114 km/h (71 mi/h). While chasing its prey in a short sprint, a cheetah starts from rest and runs 45 m in a straight line, reaching a final speed of 72 km/h. (a) Determine the cheetah's average acceleration during the short sprint, and (b) find its displacement at $t = 3.5$ s.

Solution

(a) We compute the average acceleration of the cheetah by determining the constant acceleration that would be required to travel the same distance (45 m) and reach the same final speed (72 km/h) during the sprint. To do this, we make use of the equations of one-dimensional motion with constant acceleration. Since we know the initial and final velocities, as well as the displacement achieved, the equation $v^2 = v_0^2 + 2a(\Delta x)$ is ideally suited for our problem.

$$a = \frac{v^2 - v_0^2}{2(\Delta x)} = \frac{\left[72 \frac{\text{km}}{\text{h}}\left(\frac{10^3 \text{ m}}{1 \text{ km}}\right)\left(\frac{1 \text{ h}}{3\,600 \text{ s}}\right)\right]^2 - 0}{2(45 \text{ m})} = 4.4 \text{ m/s}^2 \qquad \Diamond$$

(b) To find the displacement achieved during the first 3.5 seconds of the sprint, we make use of $\Delta x = v_0 t + \frac{1}{2}at^2$, with $v_0 = 0$ and $a = 4.4 \text{ m/s}^2$ as found above. This yields

$$\Delta x = v_0 t + \frac{1}{2}at^2 = 0 + \frac{1}{2}(4.4 \text{ m/s}^2)(3.5 \text{ s})^2 = 27 \text{ m} \qquad \Diamond$$

---

19. Runner A is initially 4.0 mi west of a flagpole and is running with a constant velocity of 6.0 mi/h due east. Runner B is initially 3.0 mi east of the flagpole and is running with a constant velocity of 5.0 mi/h due west. How far are the runners from the flagpole when they meet?

Solution

We choose a coordinate axis which has the origin at the flagpole and eastward as the positive x direction. Then, runner A is initially located at $x_{0A} = -4.0$ mi, has initial velocity $v_{0A} = +6.0$ mi/h, and zero acceleration $(a_A = 0)$. Similarly, runner B is initially located at $x_{0B} = +3.0$ mi, has initial velocity $v_{0B} = -5.0$ mi/h, and zero acceleration $(a_B = 0)$.

The kinematics equation $\Delta x = x - x_0 = v_0 t + \frac{1}{2}at^2$ may be used to obtain each runner's position at time t. This gives

Runner A: $x_A - (-4.0 \text{ mi}) = (6.0 \text{ mi/h})t + 0$ or $x_A = -4.0 \text{ mi} + (6.0 \text{ mi/h})t$

Runner B: $x_B - (+3.0 \text{ mi}) = (-5.0 \text{ mi/h})t + 0$ or $x_B = 3.0 \text{ mi} - (5.0 \text{ mi/h})t$

When the two runners meet, their x-coordinates will be the same $(x_A = x_B)$. Requiring that this be true and solving the resulting equation for t will give the elapsed time when they meet:

$x_A = x_B$ $\Rightarrow$ $-4.0 \text{ mi} + (6.0 \text{ mi/h})t = 3.0 \text{ mi} - (5.0 \text{ mi/h})t$

or $(6.0 \text{ mi/h} + 5.0 \text{ mi/h})t = 3.0 \text{ mi} + 4.0 \text{ mi}$

and $t = \dfrac{7.0 \text{ mi}}{11 \text{ mi/h}} = 7.0/11 \text{ h}$

At this time, $x_A = -4.0 \text{ mi} + (6.0 \text{ mi/h})(7.0/11 \text{ h}) = -0.18 \text{ mi}$

and $x_B = 3.0 \text{ mi} - (5.0 \text{ mi/h})(7.0/11 \text{ h}) = -0.18 \text{ mi}$

Thus, the runners meet 0.18 mi west of the flagpole. ◊

25. A steam catapult launches a jet aircraft from the aircraft carrier *John C. Stennis*, giving it a speed of 175 mi/h in 2.50 s. (a) Find the average acceleration of the plane. (b) Assuming the acceleration is constant, find the distance the plane moves.

Solution

(a) The plane starts from rest $(v_0 = 0)$ and reaches a velocity of $v = 175$ mi/h in an elapsed time of $\Delta t = 2.50$ s.

The average acceleration over a time interval is defined as

$$\bar{a} = \frac{\Delta v}{\Delta t} = \frac{v - v_0}{\Delta t}$$

Thus, the average acceleration of this plane during launch is

$$\bar{a} = \frac{175 \text{ mi/h} - 0}{2.50 \text{ s}} = 70.0 \frac{\text{mi}}{\text{h} \cdot \text{s}}$$ ◊

or $$\bar{a} = 70.0 \frac{\text{mi}}{\text{h} \cdot \text{s}}\left(\frac{1609 \text{ m}}{1 \text{ mi}}\right)\left(\frac{1 \text{ h}}{3600 \text{ s}}\right) = 31.3 \text{ m/s}^2$$ ◊

and $$\bar{a} = 31.3 \frac{\text{m}}{\text{s}^2}\left(\frac{1\, g}{9.80 \text{ m/s}^2}\right) = 3.19\, g$$ ◊

(b) If the acceleration is constant, the displacement of the plane in time Δt is given by $\Delta x = v_0(\Delta t) + \frac{1}{2}a(\Delta t)^2$. Assuming that the plane's acceleration is constant during launch, the distance it moves in the process is then

$$\Delta x = 0 + \frac{1}{2}\left(31.3 \frac{\text{m}}{\text{s}^2}\right)(2.50 \text{ s})^2 = 97.8 \text{ m}$$ ◊

or $$\Delta x = 97.8 \text{ m}\left(\frac{3.281 \text{ ft}}{1 \text{ m}}\right) = 321 \text{ ft}$$ ◊

29. A truck covers 40.0 m in 8.50 s while uniformly slowing down to a final velocity of 2.80 m/s. (a) Find the truck's original speed. (b) Find its acceleration.

Solution

(a) We determine the original speed of the truck by using the fact that the average velocity is $\overline{v} = (v_f + v_0)/2$ for any time interval during which the acceleration is constant. Then, from the defining equation for average velocity ($\overline{v} = \Delta x/\Delta t$), we find $\overline{v} = (v_f + v_0)/2 = \Delta x/\Delta t$, or

$$v_0 = 2\left(\frac{\Delta x}{\Delta t}\right) - v_f = 2\left(\frac{40.0 \text{ m}}{8.50 \text{ s}}\right) - 2.80 \text{ m/s} = 6.61 \text{ m/s} \qquad \Diamond$$

(b) Now that both the initial and final speed of the truck are known, its constant acceleration is found from $a = \Delta v/\Delta t$ as

$$a = \frac{v_f - v_0}{\Delta t} = \frac{2.80 \text{ m/s} - 6.61 \text{ m/s}}{8.50 \text{ s}} = -0.448 \text{ m/s}^2 \qquad \Diamond$$

---

35. Speedy Sue, driving at 30.0 m/s, enters a one-lane tunnel. She then observes a slow-moving van 155 m ahead traveling at 5.00 m/s. Sue applies her brakes but can accelerate only at -2.00 m/s^2 because the road is wet. Will there be a collision? State how you decide. If yes, determine how far into the tunnel and at what time the collision occurs. If no, determine the distance of closest approach between Sue's car and the van.

Solution

Choose $x = 0$ and $t = 0$ at the location of Sue's car when she first spots the van and applies the brakes, and take the positive direction to be the direction the car is traveling. Then, the initial conditions for Sue's car are $x_{0S} = 0$, and $v_{0S} = +30.0 \text{ m/s}$. The initial conditions for the van are $x_{0V} = +155 \text{ m}$, and $v_{0V} = +5.00 \text{ m/s}$. For times $t > 0$, the constant accelerations of the two vehicles are $a_S = -2.00 \text{ m/s}^2$ and $a_V = 0$.

These initial conditions and the kinematics equation $\Delta x = x - x_0 = v_0 t + \frac{1}{2}at^2$ will give the positions of each vehicle as functions of time:

Sue's Car: $$x_S - 0 = (30.0 \text{ m/s})t + \frac{1}{2}(-2.00 \text{ m/s}^2)t^2$$

or $$x_S = (30.0 \text{ m/s})t - (1.00 \text{ m/s}^2)t^2$$

Van: $$x_V - 155 \text{ m} = (5.00 \text{ m/s})t + \frac{1}{2}(0)t^2$$

or $$x_V = 155 \text{ m} + (5.00 \text{ m/s})t$$

If a collision is to occur, the two vehicles must be at the same location $(x_S = x_V)$ at some time. Thus, we test for a collision by equating the equations for the x coordinates and see if the resulting equation has any real solutions.

$x_S = x_V$: $$(30.0\text{ m/s})t - (1.00\text{ m/s}^2)t^2 = 155\text{ m} + (5.00\text{ m/s})t$$

or $$(1.00\text{ m/s}^2)t^2 + (-25.0\text{ m/s})t + 155\text{ m} = 0$$

The quadratic formula then gives

$$t = \frac{-(-25.0\text{ m/s}) \pm \sqrt{(-25.0\text{ m/s})^2 - 4(1.00\text{ m/s}^2)(155\text{ m})}}{2(1.00\text{ m/s}^2)}$$

or $t = 13.6\text{ s}$ (using upper sign) and $t = 11.4\text{ s}$ (using lower sign)

These solutions are real, not imaginary, so a collision will occur! ◊

The smaller of the above solutions is the actual collision time. The larger solution tells when the van would again come even with the car (and start to pull ahead) if the two vehicles could pass harmlessly through each other. The location of the collision (measured from the start of the tunnel where Sue first spotted the van) is

$$x_S\big|_{t=11.4\text{ s}} = x_V\big|_{t=11.4\text{ s}} = 155\text{ m} + (5.00\text{ m/s})(11.4\text{ s}) = 212\text{ m}$$ ◊

---

39. A car starts from rest and travels for 5.0 s with a uniform acceleration of +1.5 m/s^2. The driver then applies the brakes, causing a uniform acceleration of –2.0 m/s^2. If the brakes are applied for 3.0 s, (a) how fast is the car going at the end of the braking period, and (b) how far has the car gone?

Solution

During the first acceleration period of duration $\Delta t_1 = 5.0\text{ s}$, the car starts from rest $(v_{01} = 0)$ and accelerates at $a_1 = +1.5\text{ m/s}^2$. The velocity of the car at the end of this period is

$$v_{f1} = v_{01} + a_1(\Delta t_1) = 0 + (1.5\text{ m/s}^2)(5.0\text{ s}) = 7.5\text{ m/s}$$

This is also the velocity of the car at the start of the braking period $(v_{02} = v_{f1} = 7.5\text{ m/s})$.

(a) During the braking period, the acceleration is $a_2 = -2.0\text{ m/s}^2$ and $\Delta t_2 = 3.0\text{ s}$. The velocity at the end of this period will be

$$v_{f2} = v_{02} + a_2(\Delta t_2) = 7.5\text{ m/s} + (-2.0\text{ m/s}^2)(3.0\text{ s}) = 1.5\text{ m/s}$$ ◊

(b) The total distance traveled during the two periods is

$$\Delta x_{\text{total}} = \Delta x_1 + \Delta x_2 = \bar{v}_1 \Delta t_1 + \bar{v}_2 \Delta t_2 = \left(\frac{v_{f1} + v_{01}}{2} \right) \Delta t_1 + \left(\frac{v_{f2} + v_{02}}{2} \right) \Delta t_2$$

or $$\Delta x_{\text{total}} = \left(\frac{7.5 \text{ m/s} + 0}{2} \right)(5.0 \text{ s}) + \left(\frac{1.5 \text{ m/s} + 7.5 \text{ m/s}}{2} \right)(3.0 \text{ s}) = 32 \text{ m}$$ ◊

---

47. A certain freely falling object, released from rest, requires 1.50 s to travel the last 30.0 m before it hits the ground. (a) Find the velocity of the object when it is 30.0 m above the ground. (b) Find the total distance the object travels during the fall.

Solution

(a) The initial velocity for the last interval of the fall ($\Delta y = -30.0$ m, $\Delta t = 1.50$ s) of this object may be determined by use of the kinematics equation $\Delta y = v_0(\Delta t) + \frac{1}{2}a(\Delta t)^2$. The constant acceleration of the freely falling body is $a = -g = -9.80 \text{ m/s}^2$, so we find

$$v_0 = \frac{\Delta y - \frac{1}{2}a(\Delta t)^2}{(\Delta t)} = \frac{-30.0 \text{ m} - \frac{1}{2}(-9.80 \text{ m/s}^2)(1.50 \text{ s})^2}{1.50 \text{ s}} = -12.7 \text{ m/s}$$ ◊

(b) This downward velocity ($v = -12.7 \text{ m/s}$) was attained by the falling body after starting from rest ($v_0 = 0$) and undergoing a displacement $\Delta y = -d$ before reaching the point 30.0 m above ground. We can determine the distance d by applying the kinematics equation $v^2 = v_0^2 + 2a(\Delta y)$ to this initial interval of the fall. This gives $\Delta y = (v^2 - v_0^2)/2a$, or $-d = (v^2 - v_0^2)/2(-g)$, and yields

$$d = \frac{(-12.7 \text{ m/s})^2 - 0}{2(9.80 \text{ m/s}^2)} = 8.23 \text{ m}$$

The total distance the object travels during the fall is then

$$h = d + 30.0 \text{ m} = 8.23 \text{ m} + 30.0 \text{ m} = 38.2 \text{ m}$$ ◊

---

53. A model rocket is launched straight upward with an initial speed of 50.0 m/s. It accelerates with a constant upward acceleration of $2.00\ \text{m/s}^2$ until its engines stop at an altitude of 150 m. (a) What can you say about the motion of the rocket after its engines stop? (b) What is the maximum height reached by the rocket? (c) How long after liftoff does the rocket reach its maximum height? (d) How long is the rocket in the air?

Solution

(a) After the engines stop, the rocket is a freely falling body. It continues upward for a while, slowing under the influence of gravity until it comes to rest momentarily at its maximum altitude. It then falls back to Earth, gaining speed as it falls. ◊

(b) While the engines are running, the rocket undergoes a displacement of $\Delta y_1 = +150$ m, has an acceleration of $a_1 = +2.00\ \text{m/s}^2$, and had an initial velocity of $v_{01} = +50.0$ m/s. The velocity of the rocket (v_{f1}) when the engines shut off may be found from $v^2 = v_0^2 + 2a(\Delta y)$ as

$$v_{f1} = +\sqrt{v_{01}^2 + 2a_1(\Delta y_1)} = +\sqrt{(50.0\ \text{m/s})^2 + 2(+2.00\ \text{m/s}^2)(150\ \text{m})} = 55.7\ \text{m/s}$$

This is also the initial velocity $(v_{02} = v_{f1})$ for the coasting phase of the rocket, which ends when the rocket comes to rest momentarily $(v_{f2} = 0)$ at the maximum altitude. During the coasting phase, the rocket has acceleration $a_2 = -g$, so $v^2 = v_0^2 + 2a(\Delta y)$ gives the displacement while coasting as

$$\Delta y_2 = \frac{v_{f2}^2 - v_{02}^2}{2(-g)} = \frac{0 - (55.7\ \text{m/s})^2}{2(-9.80\ \text{m/s}^2)} = 158\ \text{m}$$

The maximum height reached is $h_{\text{max}} = \Delta y_1 + \Delta y_2 = 150\ \text{m} + 158\ \text{m} = 308\ \text{m}$ ◊

(c) The time that the engines ran is given by

$$\Delta t_1 = \frac{\Delta y_1}{\bar{v}_1} = \frac{\Delta y_1}{(v_{f1} + v_{01})/2} = \frac{2(150\ \text{m})}{55.7\ \text{m/s} + 50.0\ \text{m/s}} = 2.84\ \text{s}$$

Similarly, the duration of the coasting phase of the trip is

$$\Delta t_2 = \frac{\Delta y_2}{\bar{v}_2} = \frac{\Delta y_2}{(v_{f2} + v_{02})/2} = \frac{2(158\ \text{m})}{0 + 55.7\ \text{m/s}} = 5.67\ \text{s}$$

So the total time the rocket takes to reach the maximum height is

$$\Delta t_{\text{up}} = \Delta t_1 + \Delta t_2 = 2.84\ \text{s} + 5.67\ \text{s} = 8.51\ \text{s}$$ ◊

(d) The time required for the rocket to fall to the ground, starting from rest $(v_{03} = 0)$, at an attitude of 308 m $(\text{so } \Delta y_3 = -308 \text{ m})$, with acceleration $a_3 = -g$, is given by $\Delta y_3 = v_{03}(\Delta t_{\text{down}}) + \frac{1}{2}a_3(\Delta t_{\text{down}})^2$ as

$$-308 \text{ m} = 0 + \frac{1}{2}(-9.80 \text{ m/s}^2)(\Delta t_{\text{down}})^2$$

or $$\Delta t_{\text{down}} = \sqrt{\frac{2(-308 \text{ m})}{-9.80 \text{ m/s}^2}} = 7.93 \text{ s}$$

The total time the rocket is in the air is

$$\Delta t_{\text{total}} = \Delta t_{\text{up}} + \Delta t_{\text{down}} = 8.51 \text{ s} + 7.93 \text{ s} = 16.4 \text{ s}$$ ◊

---

64. To pass a physical education class at a university, a student must run 1.0 mi in 12 min. After running for 10 min, she still has 500 yd to go. If her maximum acceleration is 0.15 m/s^2, can she make it? If the answer is no, determine what acceleration she would need to be successful.

Solution

The student has run a distance of

$$\Delta x_1 = 1.0 \text{ mi} - 500 \text{ yd} = 5\,280 \text{ ft} - 1\,500 \text{ ft} = 3\,780 \text{ ft}$$

during the first 10 minutes. Her average speed during this period, and her initial speed for the last 500 yd dash, is given by

$$v_{0,2} = \overline{v}_1 = \frac{\Delta x_1}{\Delta t_1} = \frac{3\,780 \text{ ft}}{10 \text{ min}}\left(\frac{1 \text{ min}}{60 \text{ s}}\right) = 6.3 \text{ ft/s}$$

With an initial speed of $v_0 = 6.3$ ft/s, the minimum constant acceleration that would be needed to complete the last 500 yards (1 500 ft) in the remaining 2.0 min (120 s) of her allotted time is found from $\Delta x = v_i t + \frac{1}{2}at^2$ as

$$a_{\text{min}} = \frac{2[\Delta x - v_0 t]}{t^2} = \frac{2[1\,500 \text{ ft} - (6.3 \text{ ft/s})(120 \text{ s})]}{(120 \text{ s})^2} = 0.103 \text{ ft/s}^2$$

or $$a_{\text{min}} = \left(0.103 \frac{\text{ft}}{\text{s}^2}\right)\left(\frac{1 \text{ m}}{3.281 \text{ ft}}\right) = 0.031 \text{ m/s}^2$$

Since this acceleration is considerably smaller than the acceleration of 0.15 m/s^2 that she is capable of producing, she should be able to easily meet the requirement of running 1.0 mile in 12 minutes. ◊

---

67. You drop a ball from a window on an upper floor of a building, and it is caught by a friend on the ground when the ball is moving with speed v_f. You now repeat the drop, but you have a friend on the street below throw another ball upward at speed v_f exactly at the same time that you drop your ball from the window. The two balls are initially separated by 28.7 m. (a) At what time do they pass each other? (b) At what location do they pass each other relative the window?

Solution

(a) The velocity the first ball will have when it reaches the ground (starting from rest at a height of 28.7 m) is found by using the kinematics equation $v^2 = v_0^2 + 2a(\Delta y)$ to be

$$v_f = -\sqrt{v_0^2 + 2a(\Delta y)} = -\sqrt{0 + 2(-9.80\ \text{m/s}^2)(-28.7\ \text{m})} = -23.7\ \text{m/s}$$

When the drop is repeated, we take $t = 0$ at the instant the balls are released. The initial conditions for the two balls are then:

Ball 1: $y_{0,1} = +28.7$ m and $v_{0,1} = 0$ Ball 2: $y_{0,2} = 0$ and $v_{0,2} = +23.7\ \text{m/s}$

Each ball will have a constant acceleration of $a = -g$, and the y-coordinate of each ball at any time $t > 0$ is given by $y = y_0 + v_0 t + \frac{1}{2}at^2$ as

Ball 1: $y_1 = +28.7\ \text{m} + 0 + \frac{1}{2}(-g)t^2$ or $y_1 = +28.7\ \text{m} - \left(\frac{g}{2}\right)t^2$

Ball 2: $y_2 = 0 + (+23.7\ \text{m/s})t + \frac{1}{2}(-g)t^2$ or $y_2 = (23.7\ \text{m/s})t - \left(\frac{g}{2}\right)t^2$

When the balls meet, their y-coordinates must be equal, so requiring $y_2 = y_1$ gives the elapsed time when they meet as

$$(23.7\ \text{m/s})t - \left(\frac{g}{2}\right)t^2 = 28.7\ \text{m} - \left(\frac{g}{2}\right)t^2 \quad \text{or} \quad t = \frac{28.7\ \text{m}}{23.7\ \text{m/s}} = 1.21\ \text{s} \quad \lozenge$$

(b) The distance the first ball drops below window level before they meet is

$$d = |\Delta y_1| = \left|v_{0,1}t + \frac{1}{2}at^2\right| = \left|0 + \frac{1}{2}(-9.80\ \text{m/s}^2)(1.21\ \text{s})^2\right| = 7.17\ \text{m} \quad \lozenge$$

71. An ice sled powered by a rocket engine starts from rest on a large frozen lake and accelerates at $+40\ \text{ft/s}^2$. After some time t_1, the rocket engine is shut down and the sled moves with constant velocity v for a time t_2. If the total distance traveled by the sled is 17 500 ft and the total time is $90\ \text{s}$, find (a) the times t_1 and t_2 and (b) the velocity v. At the 17 500-ft mark, the sled begins to accelerate at $-20\ \text{ft/s}^2$. (c) What is the final position of the sled when it comes to rest? (d) How long does it take to come to rest?

Solution

(a) The distance the sled travels while the engine is operating for time t_1 is given by $\Delta x = v_0 t + \frac{1}{2}at^2$ as

$$\Delta x_1 = 0 + \frac{1}{2}\left(40\ \text{ft/s}^2\right)t_1^2 \qquad \text{or} \qquad \Delta x_1 = \left(20\ \text{ft/s}^2\right)t_1^2 \qquad \textbf{[1]}$$

The velocity the sled reaches before the engine shuts down, and its constant velocity during the interval t_2, is given by $v = v_0 + at$ as

$$v = 0 + \left(40\ \text{ft/s}^2\right)t_1 \qquad \text{or} \qquad v = \left(40\ \text{ft/s}^2\right)t_1 \qquad \textbf{[2]}$$

The distance the sled travels at constant velocity v during the interval t_2 is

$$\Delta x_2 = vt_2 = \left[\left(40\ \text{ft/s}^2\right)t_1\right]t_2 \qquad \text{or} \qquad \Delta x_2 = \left(40\ \text{ft/s}^2\right)t_1 t_2 \qquad \textbf{[3]}$$

We also know that $\Delta x_1 + \Delta x_2 = 17\,500\ \text{ft}$ and $t_1 + t_2 = 90\ \text{s}$, or $t_2 = 90\ \text{s} - t_1$. Using this information, along with Equations [1] and [3], we have

$$\Delta x_1 + \Delta x_2 = \left(20\ \text{ft/s}^2\right)t_1^2 + \left(40\ \text{ft/s}^2\right)t_1\left(90\ \text{s} - t_1\right) = 17\,500\ \text{ft}$$

which reduces to $t_1^2 + (-180\ \text{s})t_1 + 875\ \text{s}^2 = 0$. Using the quadratic formula to solve for t_1 gives

$$t_1 = \frac{-(-180\ \text{s}) \pm \sqrt{(-180\ \text{s})^2 - 4(1)\left(875\ \text{s}^2\right)}}{2}$$

Using the upper sign yields $t_1 = 175\ \text{s}$ and $t_2 = 90\ \text{s} - t_1 = -85\ \text{s}$, while using the lower sign gives $t_1 = 5.0\ \text{s}$ and $t_2 = +85\ \text{s}$. Since the time t_2 must be positive, we see that the valid solutions for the times are $t_1 = 5.0\ \text{s}$ and $t_2 = 85\ \text{s}$. ◊

(b) Equation [2] gives the velocity of the sled when the engine shuts down as

$$v = \left(40\ \text{ft/s}^2\right)t_1 = \left(40\ \text{ft/s}^2\right)(5.0\ \text{s}) = 200\ \text{ft/s} \qquad ◊$$

(c) The distance the sled travels during its deceleration phase is given by $v_f^2 = v_0^2 + 2a(\Delta x)$ as

$$\Delta x_3 = \frac{v_f^2 - v_0^2}{2a_3} = \frac{0 - (200\ \text{ft/s})^2}{2\left(-20\ \text{ft/s}^2\right)} = 1\,000\ \text{ft}$$

The total distance the sled travels from its initial starting point is

$$\Delta x_{\text{total}} = (\Delta x_1 + \Delta x_2) + \Delta x_3 = (17\,500 \text{ ft}) + 1\,000 \text{ ft} = 18\,500 \text{ ft} \qquad \Diamond$$

(d) The time required for the sled to come to rest, after it starts decelerating, is given by $v_f = v_0 + at$ as

$$t_3 = \frac{v_f - v}{a_3} = \frac{0 - 200 \text{ ft/s}}{-20 \text{ ft/s}^2} = 10 \text{ s} \qquad \Diamond$$

The total time for the trip is $t_{\text{total}} = (t_1 + t_2) + t_3 = (90 \text{ s}) + 10 \text{ s} = 100 \text{ s}$. ◊

3

Vectors and Two-Dimensional Motion

NOTES FROM SELECTED CHAPTER SECTIONS

3.1 Vectors and Their Properties
3.2 Components of a Vector

A **scalar** has only magnitude and no direction. On the other hand, a **vector** is a physical quantity that requires the specification of both direction and magnitude. In Cartesian coordinates the **components of a vector** are the projections of the vector along the axes of a rectangular coordinate system. *A vector can be completely described by its components.*

Equal vectors have both the same magnitude and direction. When two or more vectors are **added**, they must have the same units. Two vectors that are the **negative** of each other have the same magnitude but opposite directions. **Multiplication** by a positive (negative) scalar results in a vector in the same (opposite) direction, with a magnitude equal to the product of the scalar and the magnitude of the original vector.

3.3 Displacement, Velocity, and Acceleration in Two Dimensions

Consider an object moving between two points in space. The **displacement** of the object is defined as the change in its position vector, $\Delta\vec{\mathbf{r}}$. The **average velocity** of a particle during the time interval Δt is the ratio of its displacement to the time interval for this displacement. The average velocity is a vector quantity directed along $\Delta\vec{\mathbf{r}}$. The **instantaneous velocity** $\vec{\mathbf{v}}$ is defined as the limit of the average velocity as Δt goes to zero. The direction of the instantaneous velocity vector is along a line that is tangent to the path of the particle and in the direction of motion. The **average acceleration** of an object whose velocity changes is the ratio of the net change in velocity to the time interval during which the change occurs.

In a given time interval, a particle can accelerate in several ways: the magnitude of the velocity vector (the speed) may change; the direction of the velocity vector may change, making a curved path, even though the speed is constant; or both the magnitude and direction of the velocity vector may change.

3.4 Motion in Two Dimensions

In the case of **projectile motion**, if it is assumed that air resistance is negligible and that the rotation of the Earth does not affect the motion, then the motion has the following characteristics:

- The horizontal component of velocity, v_x, remains constant because there is no horizontal component of acceleration.
- The vertical component of acceleration is equal to the acceleration due to gravity.
- The vertical component of velocity, v_y, and the displacement in the y-direction are identical to those of a freely falling body.
- Projectile motion can be described as a superposition of the two motions in the x- and y-directions. *The horizontal and vertical components of the motion of a projectile are completely independent of each other.*

Review the procedure recommended in the Suggestions, Skills, and Strategies section for solving projectile motion problems.

3.5 Relative Velocity

Observations made by observers in **different frames of reference** can be related to each other using the techniques of transformation of relative velocities.

When two objects are each moving with respect to a stationary reference frame (e.g., the Earth), each moving object has a **relative velocity** with respect to the other one. You must use vector addition in order to determine the relative velocity of one object with respect to another. See the suggestions for solving relative velocity problems in the Suggestions, Skills, and Strategies section.

EQUATIONS AND CONCEPTS

The **commutative law of addition** states that when two or more vectors are added, the resultant is independent of the order of addition.

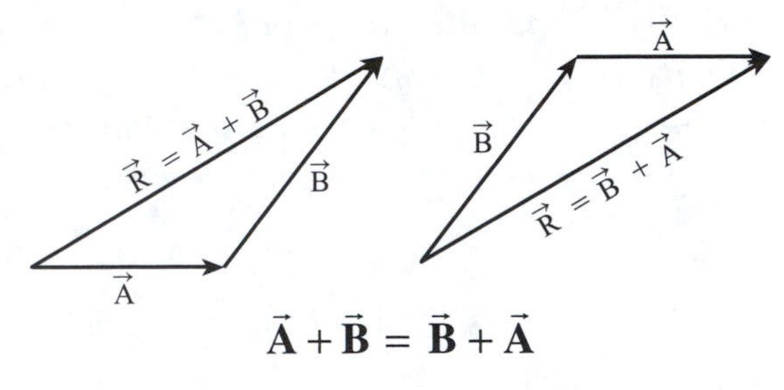

$$\vec{\mathbf{A}} + \vec{\mathbf{B}} = \vec{\mathbf{B}} + \vec{\mathbf{A}}$$

The **associative law of addition** states that when three or more vectors are added, the sum is independent of the manner in which the individual vectors are grouped.

$$\vec{\mathbf{A}} + (\vec{\mathbf{B}} + \vec{\mathbf{C}}) = (\vec{\mathbf{A}} + \vec{\mathbf{B}}) + \vec{\mathbf{C}}$$

In the **graphical or geometric method of vector addition**, the vectors to be added are represented by arrows connected head-to-tail *in any order* and the **resultant or sum** is the vector that joins the tail of the first vector to the head of the last vector. *The length of each vector must correspond to the magnitude of the vector according to a chosen scale. The direction of each vector must be along a direction which makes the proper angle relative to the others. All vectors in a sum must represent the same physical quantity (i.e., have the same units).*

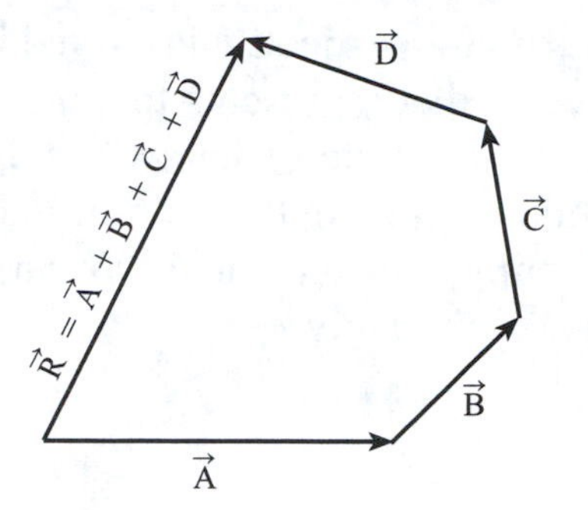

Graphical addition of four vectors.

The **operation of vector subtraction** utilizes the definition of the negative of a vector. The negative of vector $\vec{\mathbf{B}}$ is the vector that has a magnitude equal to the magnitude of $\vec{\mathbf{B}}$ but acts or points along a direction opposite the direction of $\vec{\mathbf{B}}$. Adding $(-\vec{\mathbf{B}})$ to $\vec{\mathbf{A}}$ is equivalent to subtracting $\vec{\mathbf{B}}$ from $\vec{\mathbf{A}}$.

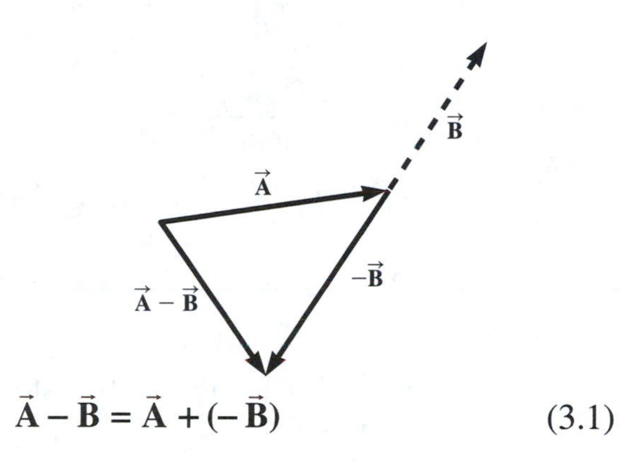

$$\vec{\mathbf{A}} - \vec{\mathbf{B}} = \vec{\mathbf{A}} + (-\vec{\mathbf{B}}) \tag{3.1}$$

The **rectangular components of a vector** are the projections of the vector onto the respective coordinate axes. The projection of $\vec{\mathbf{A}}$ onto the x-axis, A_x, is the x-component of $\vec{\mathbf{A}}$; and the projection of $\vec{\mathbf{A}}$ onto the y-axis, A_y, is the y-component of $\vec{\mathbf{A}}$. *The vector components will be positive or negative depending on the value of the angle θ which is measured counterclockwise relative to the positive x-axis.*

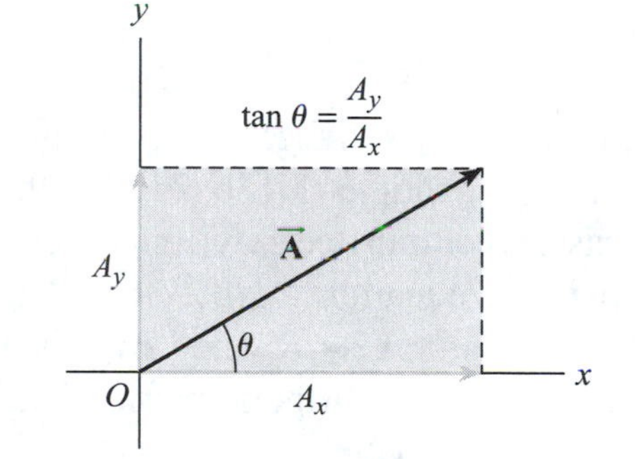

$$A_x = A\cos\theta \tag{3.2a}$$

$$A_y = A\sin\theta \tag{3.2b}$$

The **magnitude of vector $\vec{\mathbf{A}}$ and the angle** which the vector makes with the positive x-axis can be determined from the values of the x- and y-components of $\vec{\mathbf{A}}$.

$$A = \sqrt{A_x^2 + A_y^2} \tag{3.3}$$

$$\tan\theta = \frac{A_y}{A_x} \tag{3.4}$$

The **path of a projectile** is curved in the shape of a parabola as shown in the figure below. The vector that represents the initial velocity, $\vec{\mathbf{v}}_0$, makes an angle of θ_0 (called the projection angle or angle of launch) with the horizontal. In order to analyze projectile motion, you should separate the motion into two parts: the x-components (horizontal motion) and the y-components (vertical motion). Then apply the equations of constant acceleration to each part separately.

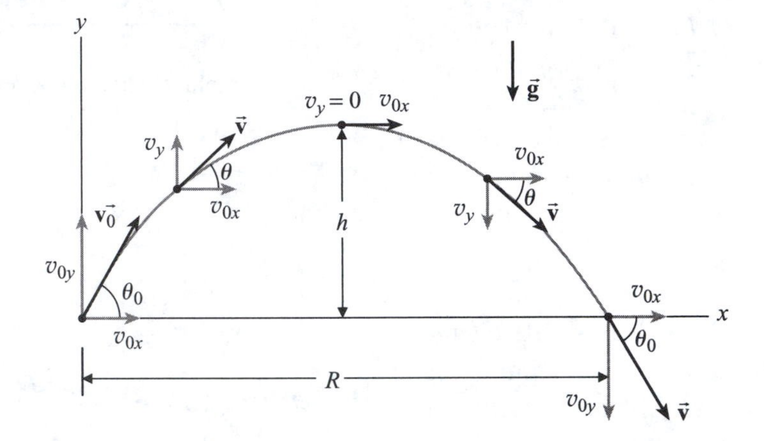

Velocity components, maximum height, and range in projectile motion

Maximum height of a projectile (h) can be found by combining Equations 3.14a and 3.14b. Remember, at maximum height, $v_y = 0$ and $\Delta y = h$.

$$h = \frac{v_0^2 \sin^2 \theta_0}{2g}$$

Range of a projectile (R) can be found by combining Equations 3.13b and 3.14b. In each case, t is eliminated between the respective pair of equations. Note that $\Delta y = h$ when $v_y = 0$ and $\Delta x = R$ when $\Delta y = 0$. You will need to use the trigonometric identity: $2 \sin\theta \cos\theta = \sin(2\theta)$.

$$R = \frac{v_0^2}{g} \sin(2\theta_0)$$

The **initial horizontal and vertical components of velocity** of a projectile depend on the magnitude of the initial velocity vector and the initial angle of launch.

$$v_{0x} = v_0 \cos\theta_0$$

$$v_{0y} = v_0 \sin\theta_0$$

The horizontal component of velocity for a projectile remains constant $(a_x = 0)$; **the vertical component** decreases uniformly with time $(a_y = -g)$.

$$v_x = v_{0x} = v_0 \cos\theta_0 = \text{constant} \qquad (3.13a)$$

$$v_y = v_0 \sin\theta_0 - gt \qquad (3.14a)$$

The ***x*- and *y*-coordinates of the position** of a projectile are functions of the elapsed

$$\Delta x = v_{0x} t = (v_0 \cos\theta_0) t \qquad (3.13b)$$

$$\Delta y = (v_0 \sin\theta_0) t - \tfrac{1}{2} g t^2 \qquad (3.14b)$$

time. *In Equations* 3.14a *and* 3.14b, *the positive direction for the vertical motion is assumed to be upward.*

The **velocity in the *y*-direction** can be related to the initial velocity and the displacement as stated in Equation 3.14c.

$$v_y^2 = (v_o \sin\theta_o)^2 - 2g\Delta y \qquad (3.14c)$$

The **magnitude and direction of the velocity vector** at any time can be determined from the values of v_x and v_y.

$$v = \sqrt{v_x^2 + v_y^2}$$

$$\theta = \tan^{-1}\left(\frac{v_y}{v_x}\right)$$

Relative velocity must be considered when both an observer and an object whose velocity is to be measured are moving with respect to a fixed frame of reference.

In Equation 3.16, A is the object whose velocity is to be measured by observer B.

$$\vec{\mathbf{v}}_{AB} = \vec{\mathbf{v}}_{AE} - \vec{\mathbf{v}}_{BE} \qquad (3.16)$$

$\vec{\mathbf{v}}_{AB}$ = velocity of A relative to B

$\vec{\mathbf{v}}_{AE}$ = velocity A relative to the fixed frame

$\vec{\mathbf{v}}_{BE}$ = velocity of B relative to the fixed frame

SUGGESTIONS, SKILLS, AND STRATEGIES

ADDITION AND SUBTRACTION OF VECTORS

When two or more vectors are to be added, the following step-by-step procedure is recommended:

1. Select a coordinate system.
2. Draw a sketch of the vectors to be added (or subtracted), with a label on each vector.
3. Find the *x*- and *y*-components of all vectors.
4. Find the algebraic sum of the *x*-components of the vectors, and the algebraic sum of the *y*-components of the vectors. These two sums are the *x*- and *y*-components of the resultant vector.
5. Use the Pythagorean theorem to find the magnitude of the resultant vector.
6. Use a suitable trigonometric function—e.g., Equation 3.4—to find the angle that the resultant vector makes with the *x*-axis.

SOLVING PROJECTILE MOTION PROBLEMS

The following procedure is recommended for solving projectile motion problems:

1. Sketch the path of the projectile on a set of coordinate axes. Include the initial velocity vector and the projection angle.

2. Resolve the initial velocity vector into *x*- and *y*-components.

3. Treat the horizontal motion (zero acceleration) and the vertical motion (acceleration $= -g$) independently.

4. Follow the techniques for solving problems with constant velocity (zero acceleration) to analyze the horizontal motion of the projectile.

5. Follow the techniques for solving problems with constant acceleration to analyze the vertical motion of the projectile.

SOLVING RELATIVE VELOCITY PROBLEMS

1. Examine the statement of the problem for phrases like ". . . the velocity of A relative to B. . . ." When a velocity is not explicitly stated as being relative to a specific observer, it is usually relative to the Earth.

2. Show the velocity vectors in a diagram; label each vector involved (usually two) with a letter that reminds you what it represents (i.e., E for Earth).

3. Assemble the three velocities you have identified into a vector equation with subscripts that represent the relationship among the velocities. For example, "the velocity of observer A relative to observer B equals the velocity of observer A relative to the Earth minus the velocity of observer B relative to the Earth" would appear as $\vec{\mathbf{v}}_{AB} = \vec{\mathbf{v}}_{AE} - \vec{\mathbf{v}}_{BE}$.

4. Solve the equation in step 3 for the unknown velocity. In the case of one-dimensional motion, this will involve one equation and one unknown; for two-dimensional motion there will be two component equations and two unknowns.

REVIEW CHECKLIST

- Understand and describe the basic properties of vectors, resolve a vector into its rectangular components, and use the rules of vector addition (including the graphical method) to find the sum (or resultant) of two or more vectors. Determine the magnitude and direction of a vector from its rectangular components.

- Sketch a typical trajectory of a particle moving in the *xy*-plane and draw vectors to illustrate the manner in which the displacement, velocity, and acceleration of the particle change with time.

- Recognize that two-dimensional motion in the *xy*-plane with constant acceleration is equivalent to two independent motions: constant velocity along the *x*-direction and constant acceleration along the *y*-direction.

- Recognize the fact that if the initial speed and initial angle of a projectile motion are known at a given point at $t = 0$, the velocity components and coordinates can be found at any later time *t*. Furthermore, one can also calculate the horizontal range *R* and maximum height *h* if v_o and θ_o are known.

- Practice the technique demonstrated in the text to solve relative velocity problems.

SOLUTIONS TO SELECTED END-OF-CHAPTER PROBLEMS

3. Vector $\vec{\mathbf{A}}$ is 3.00 units in length and points along the positive *x*-axis. Vector $\vec{\mathbf{B}}$ is 4.00 units in length and points along the negative *y*-axis. Use graphical methods to find the magnitude and direction of the vectors (a) $\vec{\mathbf{A}} + \vec{\mathbf{B}}$ and (b) $\vec{\mathbf{A}} - \vec{\mathbf{B}}$.

Solution

(a) To add vectors graphically, the first vector is drawn in the specified direction and with a length that corresponds to the magnitude of this vector in the chosen scale. For example, if the chosen scale is 1/2 inch = 1 unit, the vector drawn to represent $\vec{\mathbf{A}}$ would be 1.50 inches long and point in the positive *x*-direction. Then, to add $\vec{\mathbf{B}}$ to $\vec{\mathbf{A}}$, the vector representing $\vec{\mathbf{B}}$ is drawn, starting at the tip of $\vec{\mathbf{A}}$, and pointing in the specified direction, with a length corresponding to its magnitude on the chosen scale. In this case, the second vector starts at the tip of $\vec{\mathbf{A}}$, is drawn in the negative *y*-direction, and has a length of $(4.00 \text{ units})(0.500 \text{ inch/1 unit}) = 2.00$ inches. The resultant (vector sum) is the vector drawn from the tail of the first vector to the tip of the last vector as shown in the sketch. The magnitude and direction of the resultant is determined by measuring its length with a ruler (and multiplying by the chosen scale factor), and by measuring its direction with a protractor. In your scale drawing, you should find that

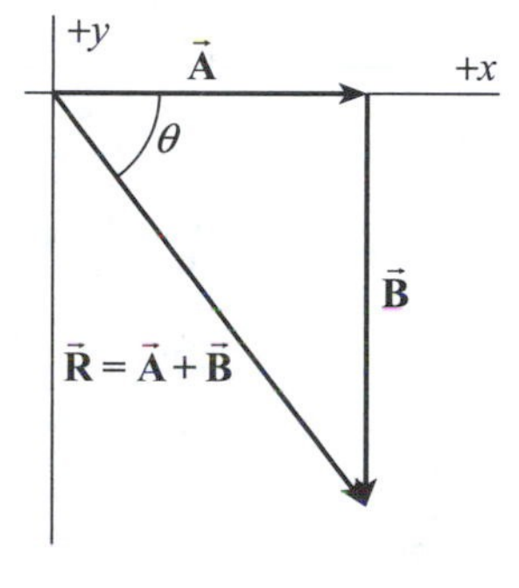

$$\vec{\mathbf{R}} = \vec{\mathbf{A}} + \vec{\mathbf{B}} = 5.0 \text{ units at } 53° \text{ below the positive } x\text{-axis} \quad \lozenge$$

(b) To subtract vector $\vec{\mathbf{B}}$ from vector $\vec{\mathbf{A}}$, realize that $\vec{\mathbf{A}} - \vec{\mathbf{B}} = \vec{\mathbf{A}} + \left(-\vec{\mathbf{B}}\right)$ and that $-\vec{\mathbf{B}}$ is a vector of the same magnitude as $\vec{\mathbf{B}}$, but in the opposite direction. The scale drawing you will need to determine $\vec{\mathbf{A}} - \vec{\mathbf{B}}$ graphically is similar to the sketch at the right. You should find that

$$\vec{\mathbf{R}} = \vec{\mathbf{A}} - \vec{\mathbf{B}} = 5.0 \text{ units at } 53° \text{ above the positive } x\text{-axis} \quad \lozenge$$

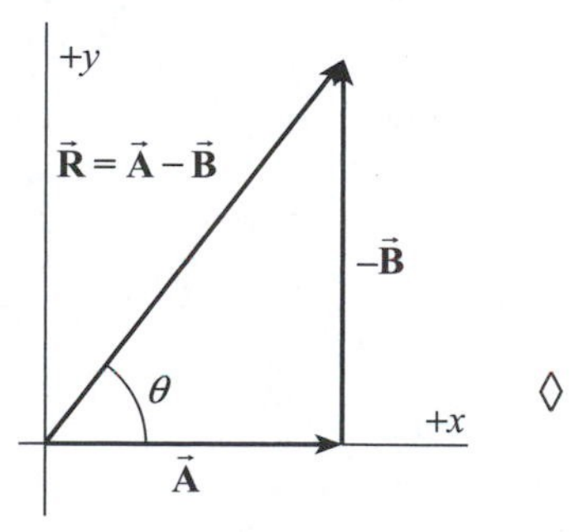

14. A hiker starts at his camp and moves the following distances while exploring his surroundings: 75.0 m north, 2.50×10^2 m east, 125 m at an angle 30.0° north of east, and 1.50×10^2 m south. (a) Find his resultant displacement from camp. (Take east as the positive x-direction and north as the positive y-direction.) (b) Would changes in the order in which the hiker makes the given displacements alter his final position? Explain.

Solution

(a) If eastward is chosen as the positive x-direction and northward is the positive y-direction, the four displacements (and their components) undergone by the hiker are as follows:

$\vec{\mathbf{A}} = 75.0$ m due north: $\quad A_x = 0$, and $A_y = +75.0$ m

$\vec{\mathbf{B}} = 2.50 \times 10^2$ m due east: $\quad B_x = +250$ m, and $B_y = 0$

$\vec{\mathbf{C}} = 125$ m at 30.0° north of east:

$$C_x = C\cos\theta = (125 \text{ m})\cos 30.0° = +108 \text{ m}$$

$$C_y = C\sin\theta = (125 \text{ m})\sin 30.0° = +62.5 \text{ m}$$

$\vec{\mathbf{D}} = 1.50 \times 10^2$ m due south: $\quad D_x = 0$, $D_y = -150$ m

The resultant displacement then has components of

$$R_x = A_x + B_x + C_x + D_x = (0 + 250 + 108 + 0) \text{ m} = 358 \text{ m}$$

$$R_y = A_y + B_y + C_y + D_y = (+75.0 + 0 + 62.5 - 150) \text{ m} = -12.5 \text{ m}$$

The magnitude and direction of the resultant displacement are given by

$$R = \sqrt{R_x^2 + R_y^2} = \sqrt{(358 \text{ m})^2 + (-12.5 \text{ m})^2} = 358 \text{ m}$$

$$\theta = \tan^{-1}\left(\frac{R_y}{R_x}\right) = \tan^{-1}\left(\frac{-12.5 \text{ m}}{358 \text{ m}}\right) = -2.00°$$

so $\vec{\mathbf{R}} = 358$ m at 2.00° south of east ◊

(b) From the calculations for R_x and R_y above, we see that the resultant will have the same components regardless of the order in which the displacements are added. Thus, the magnitude and direction of the resultant would not be changed if the hiker changed the order in which he took the four displacements. ◊

19. A commuter airplane starts from an airport and takes the route shown in Figure P3.19. It first flies to city *A* located 175 km away in a direction 30.0° north of east. Next, it flies for 150 km 20.0° west of north, to city *B*. Finally, the plane flies 190 km due west, to city *C*. Find the location of city *C* relative to the location of the starting point.

Solution

The three successive displacements undergone by the airplane are shown in Figure P3.19 as

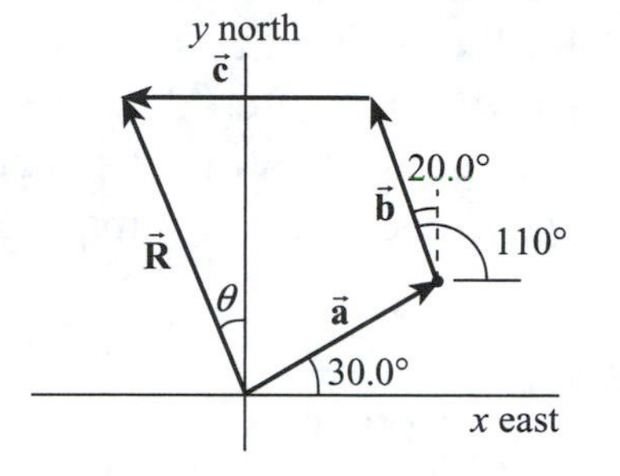

$$\vec{\mathbf{a}} = 175 \text{ km at } 30.0° \text{ north of east}$$

$$\vec{\mathbf{b}} = 150 \text{ km at } 20.0° \text{ west of north}$$

and

$$\vec{\mathbf{c}} = 190 \text{ km directed due west}$$

The displacement of city C relative to the location of the starting point is the vector sum of these three individual displacements. To compute this sum, we start by resolving the displacements $\vec{\mathbf{a}}$, $\vec{\mathbf{b}}$, and $\vec{\mathbf{c}}$ into their x-components (eastward) and y-components (northward). These components are:

$$a_x = +|\vec{\mathbf{a}}|\cos 30.0° = +(175 \text{ km})\cos 30.0 = +152 \text{ km}$$

$$a_y = +|\vec{\mathbf{a}}|\sin 30.0° = +(175 \text{ km})\sin 30.0 = +87.5 \text{ km}$$

$$b_x = -|\vec{\mathbf{b}}|\sin 20.0° = -(150 \text{ km})\sin 20.0° = -51.3 \text{ km}$$

$$b_y = +|\vec{\mathbf{b}}|\cos 20.0° = +(150 \text{ km})\cos 20.0° = +141 \text{ km}$$

$$c_x = -190 \text{ km} \quad \text{and} \quad c_y = 0$$

The components of the resultant displacement are found to be

$$R_x = a_x + b_x + c_x = (152 - 51.3 - 190) \text{ km} = -89.3 \text{ km}$$

and

$$R_y = a_y + b_y + c_y = (87.5 + 141 + 0) \text{ km} = +229 \text{ km}$$

The magnitude and direction of this resultant displacement are given by

$$R = |\vec{\mathbf{R}}| = \sqrt{R_x^2 + R_y^2} = \sqrt{(-89.3 \text{ km})^2 + (229 \text{ km})^2} = 246 \text{ km}$$

and

$$\theta = \tan^{-1}(|R_x|/R_y) = \tan^{-1}(89.3 \text{ km}/229 \text{ km}) = 21.3°$$

or the resultant displacement is $\vec{\mathbf{R}} = 246 \text{ km at } 21.3° \text{ west of north}$ ◊

23. A student stands at the edge of a cliff and throws a stone horizontally over the edge with a speed of 18.0 m/s. The cliff is 50.0 m above a flat, horizontal beach as shown in Figure P3.23. (a) What are the coordinates of the initial position of the stone? (b) What are the components of the initial velocity? (c) Write the equations for the x- and y-components of the velocity of the stone with time. (d) Write the equations for the position of the stone with time, using the coordinates shown in Figure P3.23. (e) How long after being released does the stone strike the beach below the cliff? (f) With what speed and angle of impact does the stone land?

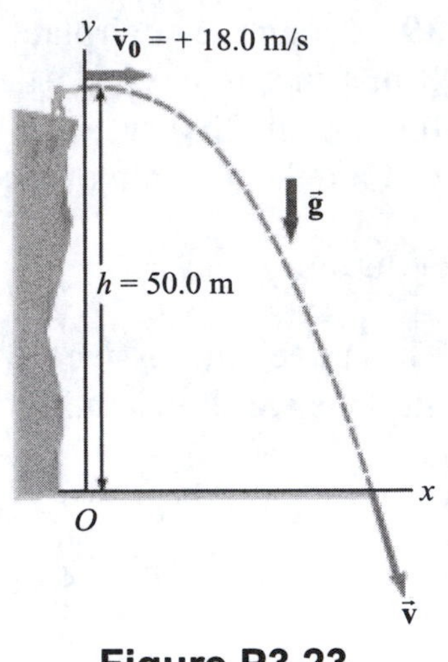

Figure P3.23

Solution

(a) With the origin chosen at point O as shown in Figure P3.23, the coordinates of the original position of the stone are $x_0 = 0$ and $y_0 = +50.0$ m. ◊

(b) The components of the initial velocity of the stone are

$$v_{0x} = +18.0 \text{ m/s} \quad \text{and} \quad v_{0y} = 0$$ ◊

(c) The components of the stone's velocity during its flight are given as functions of time by

$$v_x = v_{0x} + a_x t = 18.0 \text{ m/s} + (0)t \quad \text{or} \quad v_x = 18.0 \text{ m/s}$$ ◊

and

$$v_y = v_{0y} + a_y t = 0 + (-g)t \quad \text{or} \quad v_y = -(9.80 \text{ m/s}^2)t$$ ◊

(d) The coordinates of the stone at elapsed time t during its flight are

$$x = x_0 + v_{0x}t + \frac{1}{2}a_x t^2 = 0 + (18.0 \text{ m/s})t + \frac{1}{2}(0)t^2 \quad \text{or} \quad x = (18.0 \text{ m/s})t$$ ◊

and

$$y = y_0 + v_{0y}t + \frac{1}{2}a_y t^2 = 50.0 \text{ m} + (0)t + \frac{1}{2}(-g)t^2$$

or

$$y = 50.0 \text{ m} - (4.90 \text{ m/s}^2)t^2$$ ◊

(e) Since $\Delta y = y - y_0 = 0 - 50.0 \text{ m}$ at the instant the stone strikes the beach, we find the time of fall from $\Delta y = v_{0y}t + \frac{1}{2}a_y t^2$ with $v_{0y} = 0$:

$$t = \sqrt{\frac{2(\Delta y)}{a_y}} = \sqrt{\frac{2(-50.0 \text{ m})}{-9.80 \text{ m/s}^2}} = 3.19 \text{ s}$$ ◊

(f) At impact, $v_x = v_{0x} = 18.0$ m/s, and the vertical component of velocity is

$$v_y = v_{0y} + a_y t = 0 + (-9.80\ \text{m/s}^2)(3.19\ \text{s}) = -31.3\ \text{m/s}$$

Thus, the speed just before impact is

$$v = \sqrt{v_x^2 + v_y^2} = \sqrt{(18.0\ \text{m/s})^2 + (-31.3\ \text{m/s})^2} = 36.1\ \text{m/s} \qquad \lozenge$$

and the direction of travel is

$$\theta = \tan^{-1}\left(\frac{v_y}{v_x}\right) = \tan^{-1}\left(\frac{-31.3}{18.0}\right) = -60.1° \qquad \lozenge$$

or the velocity just before impact is

$$\vec{\mathbf{v}} = 36.1\ \text{m/s at } 60.1° \text{ below the horizontal} \qquad \lozenge$$

---

28. From the window of a building, a ball is tossed from a height y_0 above the ground with an initial velocity of 8.00 m/s and angle of 20.0° below the horizontal. It strikes the ground 3.00 s later. (a) If the base of the building is taken to be the origin of the coordinates, with upward the positive y-direction, what are the initial coordinates of the ball? (b) With the positive x-direction chosen to be out the window, find the x- and y-components of the initial velocity. (c) Find the equations for the x- and y-components of the position as functions of time. (d) How far horizontally from the base of the building does the ball strike the ground? (e) Find the height from which the ball was thrown. (f) How long does it take the ball to reach a point 10.0 m below the level of launching?

Solution

(a) When we choose the origin at ground level directly below the window, the initial coordinates of the ball are $x = 0$ and $y = y_0$. ◊

(b) The ball has an initial velocity $\vec{\mathbf{v}}_0 = 8.00$ m/s at $\theta = -20.0°$. The components of this initial velocity are:

$$v_{0x} = v_0 \cos\theta = (8.00\ \text{m/s})\cos(-20.0°) = +7.52\ \text{m/s} \qquad \lozenge$$

and

$$v_{0y} = v_0 \sin\theta = (8.00\ \text{m/s})\sin(-20.0°) = -2.74\ \text{m/s} \qquad \lozenge$$

(c) The x- and y-coordinates of the ball at time t will be

$$x = x_0 + v_{0x}t + \tfrac{1}{2}a_x t^2 = 0 + (7.52\ \text{m/s})t + \tfrac{1}{2}(0)t^2 \quad \text{or} \quad x = (7.52\ \text{m/s})t \qquad \lozenge$$

and

$$y = y_0 + v_{0y}t + \tfrac{1}{2}(-g)t^2 \quad \text{or} \quad y = y_0 + (-2.74\ \text{m/s})t - (4.90\ \text{m/s}^2)t^2 \qquad \lozenge$$

(d) When the ball strikes the ground at $t = 3.00$ s, its x-coordinate will be

$$x = (7.52 \text{ m/s})t = (7.52 \text{ m/s})(3.00 \text{ s}) = 22.6 \text{ m} \quad \Diamond$$

(e) Since, from the result of part (c), we have $y_0 = y + (2.74 \text{ m/s})t + (4.90 \text{ m/s}^2)t^2$, and we know that $y = 0$ at $t = 3.00$ s, the initial height of the ball must have been

$$y_0 = 0 + (2.74 \text{ m/s})(3.00 \text{ s}) + (4.90 \text{ m/s}^2)(3.00 \text{ s})^2 = 52.3 \text{ m} \quad \Diamond$$

(f) In part (c), the vertical displacement of the ball from its initial height is found to be given by $\Delta y = y - y_0 = (-2.74 \text{ m/s})t - (4.90 \text{ m/s}^2)t^2$. Thus, when $\Delta y = -10.0$ m, we find $(4.90 \text{ m/s}^2)t^2 + (2.74 \text{ m/s})t - 10.0 \text{ m} = 0$. Applying the quadratic formula gives solutions of

$$t = \frac{-2.74 \text{ m/s} \pm \sqrt{(2.74 \text{ m/s})^2 - 4(4.90 \text{ m/s}^2)(-10.0 \text{ m})}}{2(4.90 \text{ m/s}^2)}$$

Since the ball will be 10.0 m below the window some time after it is released ($t > 0$), we use only the positive solution and find $t = 1.18$ s. ◊

---

35. A jet airliner moving initially at 3.00×10^2 mi/h due east enters a region where the wind is blowing 1.00×10^2 mi/h in a direction 30.0° north of east. (a) Find the components of the velocity of the jet airliner relative to the air, $\vec{\mathbf{v}}_{JA}$. (b) Find the components of the velocity of the air relative to Earth, $\vec{\mathbf{v}}_{AE}$. (c) Write an equation analogous to Equation 3.16 for the velocities $\vec{\mathbf{v}}_{JA}$, $\vec{\mathbf{v}}_{AE}$, and $\vec{\mathbf{v}}_{JE}$. (d) What are the speed and direction of the aircraft relative to the ground?

Solution

(a) With eastward chosen as the positive x-direction and northward as the positive y-direction, the components of the velocity of the airliner relative to the air are given by $(\vec{\mathbf{v}}_{JA})_x = 3.00 \times 10^2$ mi/h and $(\vec{\mathbf{v}}_{JA})_y = 0$. ◊

(b) Since the wind is blowing in a direction of 30.0° north of east at a speed of 1.00×10^2 mi/h, the components of the velocity of the air relative to Earth are

$$(\vec{\mathbf{v}}_{AE})_x = (1.00 \times 10^2 \text{ mi/h})\cos 30.0° = 86.6 \text{ mi/h} \quad \Diamond$$

and

$$(\vec{\mathbf{v}}_{AE})_y = (1.00 \times 10^2 \text{ mi/h})\sin 30.0° = 50.0 \text{ mi/h} \quad \Diamond$$

(c) Note that the subscripts in Equation 3.16 in the textbook follow a very distinctive pattern. When two objects, such as cars A and B, move relative to a third object, such as the Earth, with velocities $\vec{\mathbf{v}}_{AE}$ and $\vec{\mathbf{v}}_{BE}$, respectively, the velocity of A relative to B is given by $\vec{\mathbf{v}}_{AB} = \vec{\mathbf{v}}_{AE} - \vec{\mathbf{v}}_{BE}$. In the current situation, the jet and the air both move relative to Earth with velocities $\vec{\mathbf{v}}_{JE}$ and $\vec{\mathbf{v}}_{AE}$. Following the pattern of subscripts illustrated in Equation 3.16, we see that the velocity of the jet relative to the air would be given by $\vec{\mathbf{v}}_{JA} = \vec{\mathbf{v}}_{JE} - \vec{\mathbf{v}}_{AE}$. ◊

(d) Rearranging the result of part (c) above, we find the velocity of the airliner relative to Earth to be given by $\vec{\mathbf{v}}_{JE} = \vec{\mathbf{v}}_{JA} + \vec{\mathbf{v}}_{AE}$. This velocity then has components of

$$\left(\vec{\mathbf{v}}_{JE}\right)_x = \left(\vec{\mathbf{v}}_{JA}\right)_x + \left(\vec{\mathbf{v}}_{AE}\right)_x = 3.00\times 10^2 \text{ mi/h} + 86.6 \text{ mi/h} = 387 \text{ mi/h}$$

$$\left(\vec{\mathbf{v}}_{JE}\right)_y = \left(\vec{\mathbf{v}}_{JA}\right)_y + \left(\vec{\mathbf{v}}_{AE}\right)_y = 0 + 50.0 \text{ mi/h} = 50.0 \text{ mi/h}$$

The speed of the airliner relative to Earth is the magnitude of the velocity $\vec{\mathbf{v}}_{JE}$, and is given by

$$\left|\vec{\mathbf{v}}_{JE}\right| = \sqrt{\left(\vec{\mathbf{v}}_{JE}\right)_x^2 + \left(\vec{\mathbf{v}}_{JE}\right)_y^2} = \sqrt{(387\text{ mi/h})^2 + (50.0\text{ mi/h})^2} = 390\text{ mi/h}$$ ◊

The direction of the aircraft relative to Earth is given by

$$\theta = \tan^{-1}\left(\frac{\left(\vec{\mathbf{v}}_{JE}\right)_y}{\left(\vec{\mathbf{v}}_{JE}\right)_x}\right) = \tan^{-1}\left(\frac{50.0\text{ mi/h}}{387\text{ mi/h}}\right) = 7.36°$$ ◊

The velocity of the jet relative to Earth may be written as

$$\vec{\mathbf{v}}_{JE} = 3.90\times 10^2 \text{ mi/h at } 7.36° \text{ north of east}$$ ◊

43. A bomber is flying horizontally over level terrain at a speed of 275 m/s relative to the ground and at an altitude of 3.00 km. (a) The bombardier releases one bomb. How far does the bomb travel horizontally between its release and its impact on the ground? Ignore the effects of air resistance. (b) Firing from the people on the ground suddenly incapacitates the bombardier before he can call, "Bombs away!" Consequently, the pilot maintains the plane's original course, altitude, and speed through a storm of flak. Where is the plane relative to the bomb's point of impact when the bomb hits the ground? (c) The plane has a telescopic bombsight set so that the bomb hits the target seen in the sight at the moment of release. At what angle from the vertical was the bombsight set?

Solution

(a) At the instant of its release, the bomb is moving horizontally with the velocity of the plane. Thus, its initial velocity components are $v_{0x} = v_{\text{plane}} = 275$ m/s, and $v_{0y} = 0$. With $a_y = -g$, $\Delta y = v_{0y}t + \frac{1}{2}a_y t^2$ gives the time required for the bomb to fall to the ground as $t = \sqrt{-2(\Delta y)/g}$.

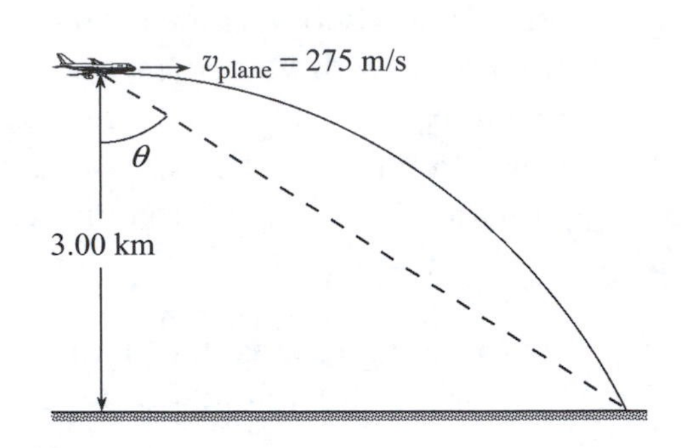

With $a_x = 0$, the horizontal distance the bomb travels during this time is

$$\Delta x = v_{0x}t + \tfrac{1}{2}a_x t^2 = v_{0x}\sqrt{-2(\Delta y)/g} + 0$$

or $$\Delta x = (275 \text{ m/s})\sqrt{\frac{-2(-3.00\times 10^3 \text{ m})}{9.80 \text{ m/s}^2}} = 6.80\times 10^3 \text{ m} = 6.80 \text{ km}$$ ◊

(b) Since the plane maintains its original course and speed, it and the bomb travel in the same horizontal direction and at the same constant horizontal speeds. As a result, the plane stays directly above the bomb as the bomb drops vertically to the ground. The plane will be 3.00 km directly above the point of impact when the bomb hits the ground. ◊

(c) As shown in the above diagram, the line of sight from the bomb's release point to point of impact makes angle θ with the vertical, where

$$\theta = \tan^{-1}\left(\frac{\text{horizontal distance bomb travels}}{\text{vertical distance bomb falls}}\right) = \tan^{-1}\left(\frac{6.80 \text{ km}}{3.00 \text{ km}}\right) = 66.2°$$ ◊

---

47. A Nordic jumper goes off a ski jump at an angle of 10.0° below the horizontal, traveling 108 m horizontally and 55.0 m vertically before landing. (a) Ignoring friction and aerodynamic effects, calculate the speed needed by the skier on leaving the ramp. (b) Olympic Nordic jumpers can make such jumps with a jump speed of 23.0 m/s, which is considerably less than the answer found in part (a). Explain how that is possible.

Solution

(a) The jumper leaves the end of the ski jump at an angle of $\theta_0 = -10.0°$ and travels a horizontal distance of $\Delta x = 108$ m while undergoing a vertical displacement of $\Delta y = -55.0$ m. Treating the jumper as a freely falling body, the components of the jumper's acceleration are $a_x = 0$ and $a_y = -g$.

If t_f is the total time of flight for the skier, the horizontal displacement at the end of the flight is $\Delta x = v_{0x}t_f = (v_0 \cos\theta)t_f$, giving $t_f = \Delta x/(v_{0x}\cos\theta)$. Substituting this into the expression for the vertical displacement, $\Delta y = v_{0y}t_f - \frac{1}{2}gt_f^2$, gives

$$\Delta y = (\cancel{v_0} \sin\theta)\left(\frac{\Delta x}{\cancel{v_0}\cos\theta}\right) - \frac{g(\Delta x)^2}{2v_0^2\cos^2\theta}$$

and solving for v_0, the speed of the skier when leaving the ramp, we find

$$v_0 = \sqrt{\frac{g(\Delta x)^2}{2(\Delta x \cdot \tan\theta - \Delta y)\cos^2\theta}} = \sqrt{\frac{(9.80 \text{ m/s}^2)(108 \text{ m})^2}{2[(108 \text{ m})\tan(-10.0°) - (-55.0 \text{ m})]\cos^2(-10.0°)}}$$

This yields $v_0 = 40.5$ m/s. ◊

(b) Olympic Nordic skiers can make this jump with a much lower initial speed by holding their arms tightly to their sides, with the body leaning forward and bent into a shape resembling an airfoil. This gives the skier some lift, allowing them to glide similar to a bird, thereby prolonging the flight and traveling farther than would be possible if they were true freely falling bodies. ◊

---

51. A rocket is launched at an angle of 53.0° above the horizontal with an initial speed of 100 m/s. The rocket moves for 3.00 s along its initial line of motion with an acceleration of 30.0 m/s^2. At this time its engines fail and the rocket proceeds to move as a projectile. Find (a) the maximum altitude reached by the rocket, (b) its total time of flight, and (c) its horizontal range.

Solution

The distance, s, moved in the first 3.00 seconds is given by

$$s = v_0 t + \frac{1}{2}at^2 = (100 \text{ m/s})(3.00 \text{ s}) + \frac{1}{2}(30.0 \text{ m/s}^2)(3.00 \text{ s})^2 = 435 \text{ m}$$

Taking the origin at the point where the rocket was launched, and upward as the positive y-direction, the coordinates of the rocket at the end of powered flight are

$$x_1 = s\cos 53.0° = 262 \text{ m} \qquad \text{and} \qquad y_1 = s\sin 53.0° = 347 \text{ m}$$

The speed of the rocket at the end of powered flight is

$$v_1 = v_0 + at = 100 \text{ m/s} + (30.0 \text{ m/s}^2)(3.00 \text{ s}) = 190 \text{ m/s}$$

so the initial velocity components for the free-fall phase of the flight are

$$v_{0x} = v_1 \cos 53.0° = 114 \text{ m/s} \qquad \text{and} \qquad v_{0y} = v_1 \sin 53.0° = 152 \text{ m/s}$$

(a) When the rocket is at maximum altitude, $v_y = 0$. The time during which the rocket continues to gain altitude at the beginning of the free-fall phase can be found from $v_y = v_{0y} + a_y t$ as

$$t_{\text{rise}} = \frac{0 - v_{0y}}{a_y} = \frac{0 - 152 \text{ m}}{-9.80 \text{ m/s}^2} = 15.5 \text{ s}$$

The vertical displacement occurring during this rise time is

$$\Delta y_{\text{rise}} = \left(\frac{v_y + v_{0y}}{2}\right) t_{\text{rise}} = \left(\frac{0 + 152 \text{ m/s}}{2}\right)(15.5 \text{ s}) = 1.18 \times 10^3 \text{ m}$$

The maximum altitude reached during the flight is then

$$H = y_1 + \Delta y_{\text{rise}} = 347 \text{ m} + 1.18 \times 10^3 \text{ m} = 1.53 \times 10^3 \text{ m}$$ ◊

(b) After reaching the top of the arc, the rocket falls 1.53×10^3 m to the ground, starting with zero vertical velocity $(v_{0y} = 0)$. The time for this fall is found from $\Delta y = v_{0y} t + \frac{1}{2} a_y t^2$ as

$$t_{\text{fall}} = \sqrt{\frac{2(\Delta y)}{a_y}} = \sqrt{\frac{2(-H)}{-g}} = \sqrt{\frac{2(-1.53 \times 10^3 \text{ m})}{-9.80 \text{ m/s}^2}} = 17.7 \text{ s}$$

Therefore, the total time of flight is

$$t = t_{\text{powered}} + t_{\text{rise}} + t_{\text{fall}} = (3.00 + 15.5 + 17.7) \text{ s} = 36.2 \text{ s}$$ ◊

(c) The duration of the free-fall phase of the flight is

$$t_2 = t_{\text{rise}} + t_{\text{fall}} = (15.5 + 17.7) \text{ s} = 33.2 \text{ s}$$

The horizontal displacement occurring during free-fall phase is

$$\Delta x_{\text{free fall}} = v_{0x} t_2 = (114 \text{ m/s})(33.2 \text{ s}) = 3.78 \times 10^3 \text{ m}$$

and the total horizontal displacement occurring during the full flight (i.e., the horizontal range of the rocket) is

$$R = x_1 + \Delta x_{\text{free fall}} = 262 \text{ m} + 3.78 \times 10^3 \text{ m} = 4.04 \times 10^3 \text{ m}$$ ◊

3.59 In a very popular lecture demonstration, a projectile is fired at a falling target as in Figure P3.59. The projectile leaves the gun at the same instant the target is dropped from rest. Assuming the gun is initially aimed at the target, show that the projectile will hit the target. (One restriction of this experiment is that the projectile must reach the target before the target strikes the floor.)

Solution

We select a coordinate system with the origin at the point where the projectile leaves the gun at time $t = 0$. At the same instant, the target is released and allowed to fall, starting from rest, under the influence of gravity. The vertical coordinates of the projectile and the target at any time while they remain in the air are given by $y = y_0 + v_{0y}t - \frac{1}{2}gt^2$ as follows.

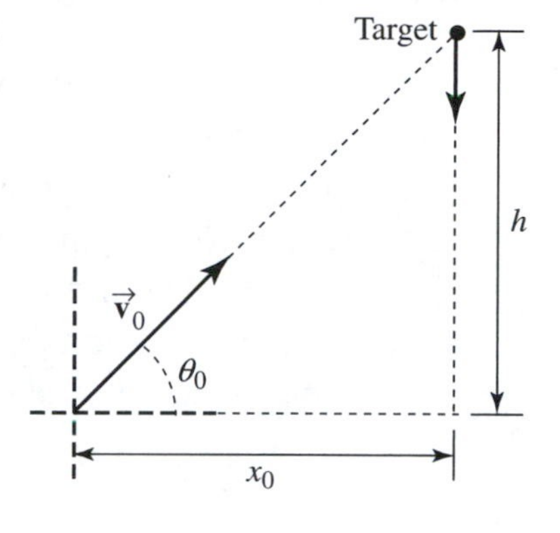

For Projectile: $y_p = 0 + (v_0 \sin\theta_0)t - (g/2)t^2$

For Target: $y_T = h + 0 - (g/2)t^2$

Since the horizontal component of the projectile's velocity is constant throughout the flight, the time when the projectile will have the same x-coordinate as the target is

$$t = \frac{x_0}{v_{0x}} = \frac{x_0}{v_0 \cos\theta_0}$$

For a collision to occur, it is necessary that $y_p = y_T$ at this time. Thus, we require that

$$(\cancel{v_0} \sin\theta_0)\left(\frac{x_0}{\cancel{v_0} \cos\theta_0}\right) - \cancel{\frac{g}{2}t^2} = h - \cancel{\frac{g}{2}t^2}$$

which reduces to

$$\tan\theta_0 = \frac{h}{x_0}$$

If this requirement is met, a collision is guaranteed to occur, provided the horizontal velocity of the projectile is large enough for the projectile to travel horizontal distance x_0 before either the projectile or the target hit the floor. Observe, from the sketch given above, that this requirement will be automatically satisfied if the gun is aimed directly at the target at the instant the projectile leaves the gun and the target is released from rest. ◊

67. A student decides to measure the muzzle velocity of a pellet shot from his gun. He points the gun horizontally. He places a target on a vertical wall a distance x away from the gun. The pellet hits the target a vertical distance y below the gun. (a) Show that the position of the pellet when traveling through the air is given by $y = Ax^2$, where A is a constant. (b) Express the constant A in terms of the initial (muzzle) velocity and the freefall acceleration. (c) If $x = 3.00$ m and $y = 0.210$ m, what is the initial speed of the pellet?

Solution

(a) We choose a coordinate system with the origin at the muzzle of the gun, with the positive x-axis oriented horizontally and pointing toward the target. The y-axis is vertical, with downward being the positive direction. Since the student holds the gun horizontally while firing it, the initial velocity components are $v_{0x} = v_0$ and $v_{0y} = 0$. Once the pellet leaves the muzzle, it is a projectile moving freely under the influence of gravity, with acceleration components of $a_x = 0$ and $a_y = +g$. Taking $t = 0$ at the instant the pellet leaves the muzzle of the gun, the coordinates of the pellet at any time t are

$$x = v_{0x}t + \tfrac{1}{2}a_x t^2 = v_0 t + 0 \qquad \text{or} \qquad x = v_0 t$$

$$\text{and} \quad y = v_{0y}t + \tfrac{1}{2}a_y t^2 = 0 + \tfrac{1}{2}a_y t^2 \qquad \text{or} \qquad y = \tfrac{1}{2}gt^2$$

Solving for the time t in the equation for the x-coordinate, and substituting the result into the equation for the y-coordinate, gives

$$y = \frac{1}{2}g\left(\frac{x}{v_0}\right)^2 = \left(\frac{g}{2v_0^2}\right)x^2 \qquad \textbf{[1]}$$

We observe that this is indeed of the form $y = Ax^2$, with A being a constant, as we wanted to verify. ◊

(b) Observing Equation [1] above, note that the vertical position of the pellet, while in the air, is given by $y = Ax^2$ with the constant A being $A = g/2v_0^2$, where g is the free-fall acceleration and v_0 is the muzzle velocity of the gun. ◊

(c) If $x = 3.00$ m and $y = 0.210$ m when the pellet hits the target, the initial speed of the pellet is

$$v_0 = \sqrt{\frac{Ax^2}{y}} = \sqrt{\frac{gx^2}{2y}} = \sqrt{\frac{(9.80 \text{ m/s}^2)(3.00 \text{ m})^2}{2(0.210 \text{ m})}} = 14.5 \text{ m/s} \qquad ◊$$

72. A dart gun is fired while being held horizontally at a height of 1.00 m above ground level, and while it is at rest relative to the ground. The dart from the gun travels a horizontal distance of 5.00 m. A college student holds the same gun in a horizontal position while sliding down a 45.0° incline at a constant speed of 2.00 m/s. How far will the dart travel if the student fires the gun when it is 1.00 m above the ground?

Solution

When the dart is fired from a stationary gun, the initial velocity components are $v_{0x} = |\vec{\mathbf{v}}_{DG}|$ and $v_{0y} = 0$, where $|\vec{\mathbf{v}}_{DG}|$ is the speed with which the dart emerges from the gun. In this case, $\Delta y = v_{0y}t + \frac{1}{2}a_y t^2$ gives the time required for the dart to reach the ground as

$$t = \sqrt{\frac{2\Delta y}{a_y}} = \sqrt{\frac{2(-1.00\ \text{m/s})}{-9.80\ \text{m/s}^2}} = 0.452\ \text{s}$$

Thus, the speed with which the dart emerges from the gun is

$$|\vec{\mathbf{v}}_{DG}| = v_{0x} = \frac{\Delta x}{t} = \frac{5.00\ \text{m}}{0.452\ \text{s}} = 11.1\ \text{m/s}$$

When the dart is fired horizontally from the moving gun, the initial velocity of the dart relative to the gun, $\vec{\mathbf{v}}_{DG}$, is the difference between the initial velocity of the dart relative to Earth, $\vec{\mathbf{v}}_{DE}$, and the velocity of the gun relative to Earth, $\vec{\mathbf{v}}_{GE}$.

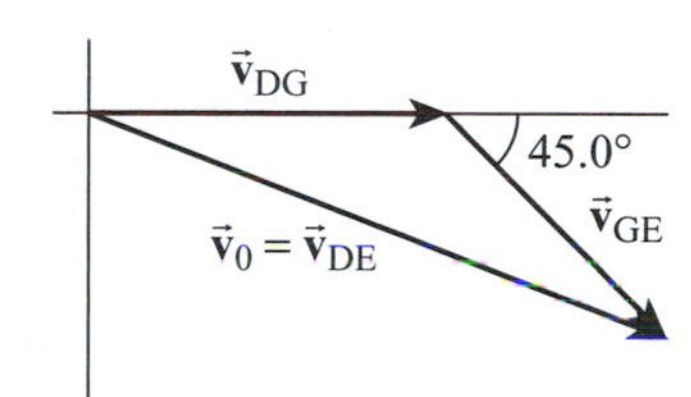

That is: $\vec{\mathbf{v}}_{DG} = \vec{\mathbf{v}}_{DE} - \vec{\mathbf{v}}_{GE}$

The initial velocity of the dart relative to Earth is then $\vec{\mathbf{v}}_0 = \vec{\mathbf{v}}_{DE} = \vec{\mathbf{v}}_{DG} + \vec{\mathbf{v}}_{GE}$, with components of

$$v_{0x} = |\vec{\mathbf{v}}_{DG}|_x + |\vec{\mathbf{v}}_{GE}|_x = 11.1\ \text{m/s} + (2.00\ \text{m/s})\cos 45.0° = 12.5\ \text{m/s}$$

and

$$v_{0y} = |\vec{\mathbf{v}}_{DG}|_y + |\vec{\mathbf{v}}_{GE}|_y = 0 - (2.00\ \text{m/s})\sin 45.0° = -1.41\ \text{m/s}$$

The vertical speed of the dart when it reaches the ground (1.00 m below its launch point) in this case is given by $v_y^2 = v_{0y}^2 + 2a_y(\Delta y)$ as

$$v_y = -\sqrt{v_{0y}^2 + 2a_y(\Delta y)} = -\sqrt{(-1.41\ \text{m/s})^2 + 2(-9.80\ \text{m/s}^2)(-1.00\ \text{m})} = -4.65\ \text{m/s}$$

The time of flight of the dart in this case is then found from $v_y = v_{0y} + a_y t$ to be

$$t = \frac{v_y - v_{0y}}{a_y} = \frac{-4.65\ \text{m/s} - (-1.41\ \text{m/s})}{-9.80\ \text{m/s}^2} = 0.331\ \text{s}$$

and the horizontal distance the dart travels during this flight is

$$\Delta x = v_{0x}t = (12.5\ \text{m/s})(0.331\ \text{s}) = 4.14\ \text{m}$$

◊

4

Laws of Motion

NOTES FROM SELECTED CHAPTER SECTIONS

4.1 Forces

Equilibrium is the condition under which the net force (vector sum of all forces) acting on an object is zero. An object in equilibrium has a zero acceleration (velocity is constant or equals zero).

Fundamental forces in nature are (listed in order of decreasing strength):

- strong nuclear force (between subatomic particles)
- electromagnetic force (between electric charges at rest or in motion)
- weak nuclear force (accompanying the process of radioactive decay)
- gravitational force (attractive forces between objects due to their mass)

Classical physics is concerned with **contact forces** (which are the result of physical contact between two or more objects) and **action-at-a-distance forces** (which act through empty space and do not involve physical contact).

4.2 Newton's First Law

Newton's first law is called the **law of inertia** and states that an object at rest will remain at rest and an object in motion will remain in motion with a constant velocity (both magnitude and direction) unless acted on by a net external force.

Mass and **weight** are two different physical quantities. The weight of a body is equal to the force of gravity acting on the body and varies with location in the Earth's gravitational field. Mass is an inherent property of a body and is a measure of the body's inertia (resistance to change in its state of motion). The SI unit of mass is the **kilogram (kg)** and the unit of weight is the **newton (N)**.

4.3 Newton's Second Law

Newton's second law, the **law of acceleration**, states that the acceleration of an object is directly proportional to the resultant (or net) force acting on it and inversely proportional to its mass. *The direction of the acceleration is in the direction of the net force.*

4.4 Newton's Third Law

Newton's third law, the **law of action-reaction**, states that when two bodies interact, the force which body "A" exerts on body "B" (the **action force**) is equal in magnitude and opposite in direction to the force which body "B" exerts on body "A" (the **reaction force**). A consequence of the third law is that forces occur in pairs. *Remember that the action force and the reaction force never cancel when Newton's second law is applied to one of the bodies because only one of these forces acts on this body while the other force acts on the other body.*

4.5 Applications of Newton's Laws

Construction of a **free-body diagram** is an important step in the application of Newton's laws of motion to solve problems involving bodies in equilibrium or accelerating under the action of external forces. The diagram should **include a labeled arrow to identify each of the external forces** acting on the body whose motion (or condition of equilibrium) is to be studied. *Forces which are the reactions to external forces must not be included.* When a system consists of more than one body or mass, you must construct a free-body diagram for each mass.

4.6 Forces of Friction

When a body is in motion either on a surface or through a viscous medium such as air or water, there is resistance to the motion because the body interacts with its surroundings. We call such resistance a force of friction. Experiments show that the frictional force arises from the nature of the two surfaces. To a good approximation, both $f_{s,\max}$ (maximum force of static friction) and f_k (force of kinetic friction) are proportional to the normal force at the interface between the two surfaces.

EQUATIONS AND CONCEPTS

A **quantitative measurement of mass** (the term used to measure inertia) can be made by comparing the accelerations that a given force will produce on different bodies. If a given force acting on a body of mass m_1 produces an acceleration a_1 and the same force acting on a body of mass m_2 produces an acceleration a_2, the ratio of the two masses equals the inverse of the ratio of the two accelerations.

$$\frac{m_1}{m_2} = \frac{a_2}{a_1}$$

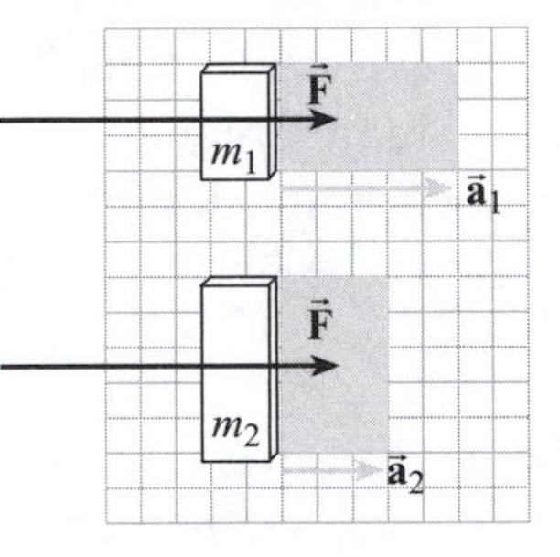

Newton's second law states that the acceleration of an object is proportional to the resultant force acting on it and inversely proportional to its mass.

$$\vec{\mathbf{a}} = \frac{\vec{\mathbf{F}}}{m}$$

$$\sum \vec{\mathbf{F}} = m\vec{\mathbf{a}} \qquad (4.1)$$

Component scalar equations are equivalent to the vector form of the equation expressing Newton's second law. *The orientation of the coordinate system can often be chosen so that the object has a nonzero acceleration along only one direction.*

$$\sum F_x = m a_x \quad (4.2)$$
$$\sum F_y = m a_y$$
$$\sum F_z = m a_z$$

The **SI unit of force** is the newton (N), defined as the force that, when acting on a 1-kg mass, produces an acceleration of 1 m/s^2. In the U.S. customary system, the unit of force is the pound. *Calculations with Equations* 4.1 *and* 4.2 *must be made using a consistent set of units for the quantities force, mass, and acceleration.*

$$1\ \text{N} \equiv 1\ \text{kg} \cdot \text{m/s}^2 \quad (4.3)$$
$$1\ \text{N} = 0.225\ \text{lb} \quad (4.4)$$

Newton's law of universal gravitation states that every particle in the universe attracts every other particle with a force that is directly proportional to the product of the masses of the particles and inversely proportional to the square of the distance between them.

$$F_g = G\frac{m_1 m_2}{r^2} \quad (4.5)$$

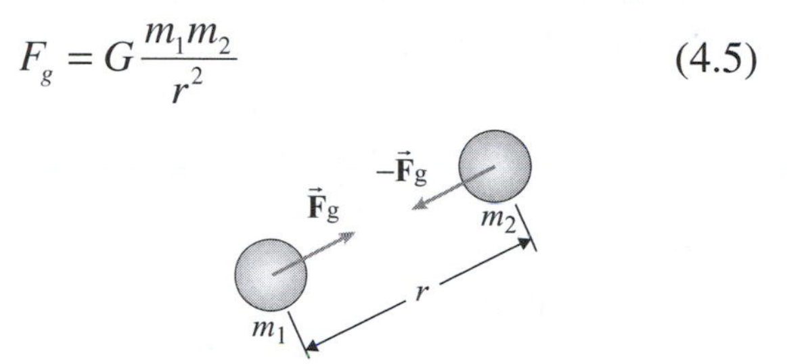

The **universal gravitational constant** is the constant of proportionality in Newton's law of universal gravitation.

$$G = 6.67 \times 10^{-11}\ \text{N} \cdot \text{m}^2/\text{kg}^2$$

Weight is not an inherent property of a body, but depends on the local value of g and varies with location.

$$w = mg \quad (4.6)$$

The **acceleration due to gravity**, g, decreases with increasing distance from the center of the Earth. In Equation 4.8, g is the value of the local gravitational field at a distance r from the center of the Earth.

$$g = G\frac{M_E}{r^2} \quad (4.8)$$

Newton's third law states that forces always occur in pairs; the force $\vec{\mathbf{F}}_{12}$ exerted by body 1 on body 2 (the action force) is equal in magnitude and opposite in direction to the force $\vec{\mathbf{F}}_{21}$ exerted by body 2 on body 1 (the reaction force). *The action and reaction forces always act on different objects; they cannot add (or cancel) to give a net force of zero.*

$$\vec{\mathbf{F}}_{12} = -\vec{\mathbf{F}}_{21}$$

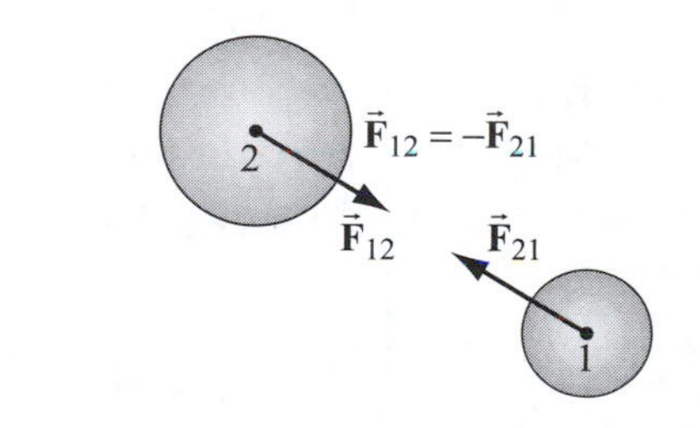

Equilibrium is a condition of rest or motion with constant velocity (magnitude and direction). *The vector sum of all forces acting on an object in equilibrium is zero.*

$$\sum \vec{\mathbf{F}} = 0 \qquad (4.9)$$

Two-dimensional equilibrium requires that the sum of all forces in the x- and y-directions must separately equal zero.

$$\sum F_x = 0 \quad \text{and} \quad \sum F_y = 0 \qquad (4.10)$$

The **force of static friction** acting on two objects whose surfaces are in contact, but *not in motion relative to each other*, cannot be greater than $\mu_s n$, where n is the magnitude of the normal (perpendicular) force between the two surfaces and μ_s is a dimensionless constant which depends on the nature of the pair of surfaces.

$$f_s \leq \mu_s n \qquad (4.11)$$

μ_s = coefficient of static friction

A **force of kinetic friction** arises when two surfaces are in relative motion. The force on each body is directed opposite the direction of motion of that body. *The coefficient of kinetic friction μ_k depends on the nature of the two surfaces.* Generally, for a given pair of surfaces, $\mu_k < \mu_s$.

$$f_k = \mu_k n \qquad (4.12)$$

μ_k = coefficient of kinetic friction

SUGGESTIONS, SKILLS, AND STRATEGIES

For problems involving the application of Newton's second law to accelerating objects and to objects in equilibrium:

1. Read the problem carefully and "picture" the description of the system of objects and forces in a simple diagram. Indicate the forces with arrows and label each force with a symbol representing the nature of the force.

2. Isolate the object of interest whose motion is to be analyzed. Draw a free-body diagram for this object; that is, a diagram showing all *external forces* acting on the object. **For systems containing more than one object, draw a separate free-body diagram for each object. Do not include forces that the object of interest exerts on other objects.**

3. Establish convenient coordinate axes for each object and find the components of the forces along these axes. In the case of an accelerating object, it is usually convenient to choose the coordinate system so that one of the axes is parallel to the direction of

motion of the object. For example, when an object is moving along an inclined plane, it might be convenient to choose the x-axis along the direction of the incline.

4. Apply Newton's second law in component form, $\sum F_x = ma_x$ and $\sum F_y = ma_y$, for each object under consideration. Remember, for an object in equilibrium, the acceleration will be zero. When forces of friction are involved. you will need to use Equation 4.12.

5. Solve the component equations for the unknowns. *Remember that you must have as many independent equations as you have unknowns in order to obtain a complete solution.* Often in solving such problems, one must also use the equations of kinematics (motion with constant acceleration) from Chapter 2 to find all the unknowns.

REVIEW CHECKLIST

- State in your own words a description of Newton's laws of motion; recall physical examples of each law, and identify the action-reaction force pairs in a multiple-body interaction problem as specified by Newton's third law.

- Express the normal force in terms of other forces acting on an object, and write out the equation which relates the coefficient of friction, force of friction, and normal force between an object and the surface on which it rests or moves.

- Apply Newton's laws of motion to various mechanical systems using the recommended procedure discussed in Section 4.5. Identify all external forces acting on each object of interest, draw a separate free-body diagram for each body of interest in the system, and apply Newton's second law, $\vec{\mathbf{F}} = m\vec{\mathbf{a}}$, in component form.

- Apply the equations of kinematics (which involve the quantities displacement, velocity, and acceleration) as described in Chapter 2 along with those methods and equations of Chapter 4 (involving mass, force, and acceleration) to the solutions of problems where both the kinematic and dynamic aspects are present.

- Be familiar with methods for solving two or more linear equations simultaneously for the unknown quantities. Recall that you must have as many independent equations as you have unknowns.

SOLUTIONS TO SELECTED END-OF-CHAPTER PROBLEMS

9. As a fish jumps vertically out of the water, assume that only two significant forces act on it: an upward force F exerted by the tail fin and the downward force due to gravity. A record Chinook salmon has a length of 1.50 m and a mass of 61.0 kg. If this fish is moving upward at 3.00 m/s as its head first breaks the surface and has an upward speed of 6.00 m/s after two-thirds of its length has left the surface, assume constant acceleration and determine (a) the salmon's acceleration and (b) the magnitude of the force F during this interval.

Solution

As the salmon approaches the surface, two forces act on it as shown in the sketch at the right if we neglect friction due to the water. The net upward force is

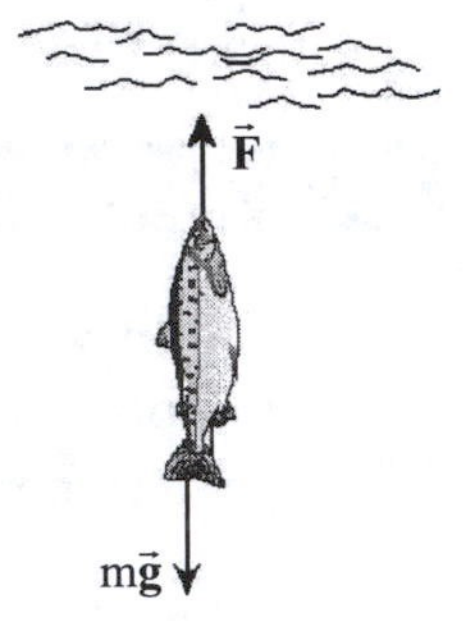

$$F_{\text{net}} = \Sigma F_y = F - mg$$

acting on the fish from the instant its head breaks the surface until it has undergone an additional upward displacement of $\Delta y = \frac{2}{3}(\textit{length}) = \frac{2}{3}(1.50\text{ m}) = 1.00\text{ m}$. As the fish undergoes this displacement, its upward acceleration is given by

$$a_y = \frac{F_{net}}{m} = \frac{F}{m} - g$$

or the magnitude of the thrust of the tail fin is

$$F = m(g + a_y)$$

(a) If the salmon's velocity increases from 3.00 m/s to 6.00 m/s during this interval, the vertical acceleration is found from $v_y^2 = v_{0y}^2 + 2a_y(\Delta y)$ as

$$a_y = \frac{v_y^2 - v_{0y}^2}{2(\Delta y)} = \frac{(6.00\text{ m/s})^2 - (3.00\text{ m/s})^2}{2(1.00\text{ m})} = 13.5\text{ m/s}^2 \quad \lozenge$$

(b) Therefore, the thrust due to the tail fin must have been

$$F = m(g + a_y) = (61.0\text{ kg})(9.80\text{ m/s}^2 + 13.5\text{ m/s}^2) = 1.42 \times 10^3\text{ N} \quad \lozenge$$

15. After falling from rest from a height of 30 m, a 0.50-kg ball rebounds upward, reaching a height of 20 m. If the contact between ball and ground lasted 2.0 ms, what average force was exerted on the ball?

Solution

When the ball starts from rest and falls freely under the influence of gravity for 30 m before reaching the ground, its speed just before contact may be found from $v_y^2 = v_{0y}^2 + 2a_y(\Delta y)$, with $v_{0y} = 0$, $a_y = -g$, and $\Delta y = -30$ m. This yields

$$v_1 = v_y = -\sqrt{v_{0y}^2 + 2a_y\Delta y} = -\sqrt{0 + 2(-9.80\ \text{m/s}^2)(-30\ \text{m})} = -24\ \text{m/s}$$

If this ball then rebounds to a height of 20 m after leaving the ground, its velocity at the instant it left the ground is found from $v_y^2 = v_{0y}^2 + 2a_y(\Delta y)$, with $v_y = 0$, $a_y = -g$, and $\Delta y = +20$ m. We find

$$v_2 = v_{0y} = +\sqrt{v_y^2 - 2a_y\Delta y} = +\sqrt{0 - 2(-9.80\ \text{m/s}^2)(+20\ \text{m})} = +20\ \text{m/s}$$

Assuming the ball was in contact with the ground for an interval $\Delta t = 2.0$ ms, the average acceleration of the ball during this interval was

$$a_{av} = \frac{v_2 - v_1}{\Delta t} = \frac{20\ \text{m/s} - (-24\ \text{m/s})}{2.0 \times 10^{-3}\ \text{s}} = +2.2 \times 10^4\ \text{m/s}^2$$

The average force exerted on the ball during this 2.0-ms time interval was then

$$F_{av} = ma_{av} = (0.50\ \text{kg})(+2.2 \times 10^4\ \text{m/s}^2) = +1.1 \times 10^4\ \text{N}$$

or

$$\vec{\mathbf{F}}_{av} = 1.1 \times 10^4\ \text{N upward}$$ ◊

23. The distance between two telephone poles is 50.0 m. When a 1.00-kg bird lands on the telephone wire midway between the poles, the wire sags 0.200 m. Draw a free-body diagram of the bird. How much tension does the bird produce in the wire? Ignore the weight of the wire.

Solution

The needed free-body diagram is:

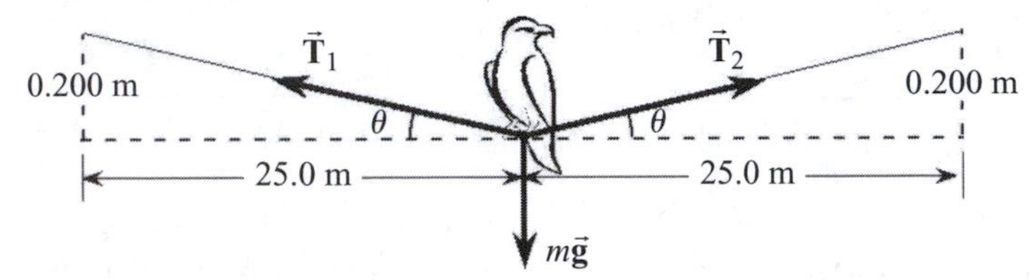

◊

Observe from the dimensions shown in the above diagram, the angle the wire makes with the horizontal is the same on each side of the bird when the bird is located at the center of the wire. This angle is

$$\theta = \tan^{-1}\left(\frac{0.200 \text{ m}}{25.0 \text{ m}}\right) = 0.458°$$

While the bird is resting on the wire, it has zero acceleration. Hence, $a_x = a_y = 0$. Considering the horizontal direction, Newton's second law gives

$$\Sigma F_x = 0 \quad \Rightarrow \quad T_2 \cos\theta - T_1 \cos\theta = 0 \quad \text{or} \quad T_2 = T_1 = T$$

where T is the common value of the tension in the wire on the two sides of the bird. Now, considering the vertical direction, Newton's second law yields

$$\Sigma F_y = 0 \quad \Rightarrow \quad T_2 \sin\theta + T_1 \sin\theta - mg = 0$$

or since $T_2 = T_1 = T$,

$$2T \sin\theta = mg$$

which gives

$$T = \frac{mg}{2\sin\theta} = \frac{(1.00 \text{ kg})(9.80 \text{ m/s}^2)}{2\sin(0.458°)} = 613 \text{ N}$$

◊

---

27. Two blocks of masses m and $2m$ are held in equilibrium on a frictionless incline as in Figure P4.27. In terms of m and θ, find (a) the magnitude of the tension T_1 in the upper cord and (b) the magnitude of the tension T_2 in the lower cord connecting the two blocks.

Solution

(a) We begin by drawing a force diagram for each of the two blocks, choosing axes that are parallel to and perpendicular to the incline, as shown below.

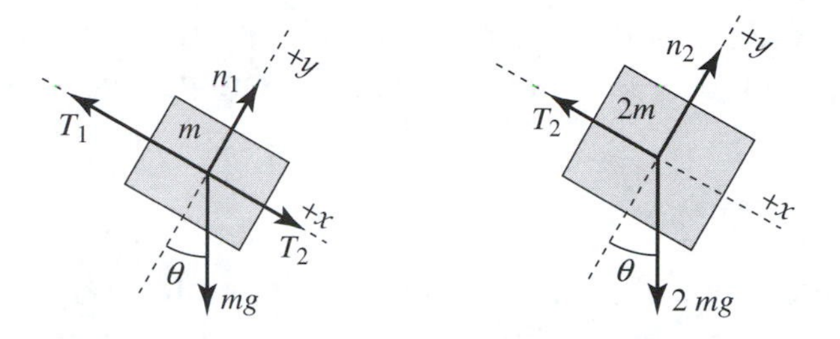

Note that the tension T_2, a force one block exerts on the other via the connecting cord, is shown acting in opposite directions on the two blocks, in agreement with Newton's third law.

Because the two blocks are in equilibrium, their accelerations are both zero. Applying Newton's second law to the motion in the x-direction for each block gives:

Block of mass m: $\Sigma F_x = T_2 + mg\sin\theta - T_1 = ma_x = 0$ or $T_1 = T_2 + mg\sin\theta$ **[1]**

Block of mass $2m$: $\Sigma F_x = 2mg\sin\theta - T_2 = (2m)a_x = 0$ or $T_2 = 2mg\sin\theta$ **[2]**

Substituting Equation [2] into Equation [1] gives

$$T_1 = 2mg\sin\theta + mg\sin\theta \quad \text{or} \quad T_1 = 3mg\sin\theta \quad \Diamond$$

(b) Equation [2] above gives the tension in the cord connecting the two blocks as

$$T_2 = 2mg\sin\theta \quad \Diamond$$

35. (a) An elevator of mass m moving upward has two forces acting on it, the upward force of tension in the cable and the downward force due to gravity. When the elevator is accelerating upward, which is greater, T or w? (b) When the elevator is moving at a constant velocity upward, which is greater, T or w? (c) When the elevator is moving upward, but the acceleration is downward, which is greater, T or w? (d) Let the elevator have a mass of 1 500 kg and an upward acceleration of $2.50\ \text{m/s}^2$. Find T. Is your answer consistent with the answer to part (a)? (e) The elevator of part (d) now moves with a constant velocity upward of 10 m/s. Find T. Is your answer consistent with your answer of part (b)? (f) Having initially moved upward with a constant velocity, the elevator begins to accelerate downward at $1.50\ \text{m/s}^2$. Find T. Is your answer consistent with your answer to part (c)?

Solution

In considering the motion of an object, it is very important to keep in mind that *the acceleration of the object is always in the direction of the net force* acting on it.

(a) When the elevator is accelerating upward, there must be a net upward force acting on it. Thus, the tension in the cable must exceed the weight of the elevator. ◊

(b) While moving at constant velocity, the elevator has zero acceleration. Hence, the net force acting on it must be zero, or the tension in the cable equals the weight of the elevator. ◊

(c) When the acceleration is directed downward, the net force must be in the downward direction, regardless of the direction of the elevator's velocity. Therefore, the weight w is greater than the tension T in this case. ◊

From Newton's second law: $\Sigma F_y = T - w = ma_y$ or $T = w + ma_y$

(d) When $a_y = +2.50\ \text{m/s}^2$, the tension in the cable is

$$T = w + (1\,500\ \text{kg})(+2.50\ \text{m/s}^2) = w + 3.75 \times 10^3\ \text{N} > w \text{ consistent with (a)}$$ ◊

or, since $w = mg = (1\,500\ \text{kg})(9.80\ \text{m/s}^2) = 1.47 \times 10^4\ \text{N}$,

$$T = 1.47 \times 10^4\ \text{N} + 3.75 \times 10^3\ \text{N} = 1.85 \times 10^4\ \text{N}$$ ◊

(e) When v is constant, $a_y = 0$ and $T = w + 0 = w = 1.47 \times 10^4\ \text{N}$, consistent with (b). ◊

(f) When $a_y = -1.50\ \text{m/s}^2$

$$T = w + (1\,500\ \text{kg})(-1.50\ \text{m/s}^2) = 1.47 \times 10^4\ \text{N} - 2.25 \times 10^3\ \text{N}$$

or $T = 1.25 \times 10^4\ \text{N} < w$ consistent with (c) ◊

43. Consider a large truck carrying a heavy load, such as steel beams. A significant hazard for the driver is that the load may slide forward, crushing the cab, if the truck stops suddenly in an accident or even in braking. Assume, for example, that a 10 000-kg load sits on the flat bed of a 20 000-kg truck moving at 12.0 m/s. Assume the load is not tied down to the truck, but has a coefficient of static friction of 0.500 with the truck bed. (a) Calculate the minimum stopping distance for which the load will not slide forward relative to the truck. (b) Is any piece of data unnecessary for the solution?

Solution

(a) The only horizontal force that can act on the load is the friction force exerted by the bed of the truck. If the load is not slipping, this will be a static friction force, f_s. When the truck slows for any reason, the load tends to slide forward on the bed, and the friction force will be directed toward the rear to oppose this motion.

The only vertical forces acting on the load are the upward normal force and the downward force of gravity. Since the load has no vertical acceleration, Newton's second law says that

$$\Sigma F_y = n - m_{\text{load}}g = m_{\text{load}}a_y = 0 \quad \text{or} \quad n = m_{\text{load}}g$$

Thus, the maximum friction force the bed can exert on the load has magnitude

$$(f_s)_{\text{max}} = \mu_s n = \mu_s m_{\text{load}}g$$

Now, applying Newton's second law in the horizontal direction when the truck is slowing, at the maximum rate without causing the load to slip, gives

$$\Sigma F_x = -(f_s)_{\text{max}} = m_{\text{load}}(a_x)_{\text{max}} \quad \text{or} \quad (a_x)_{\text{max}} = -\frac{(f_s)_{\text{max}}}{m_{\text{load}}} = -\frac{\mu_s \cancel{m_{\text{load}}} g}{\cancel{m_{\text{load}}}} = -\mu_s g$$

The minimum distance required for the truck to safely stop is given by $v_x^2 = v_{0x}^2 + 2a_x(\Delta x)$ as

$$(\Delta x)_{\text{min}} = \frac{v_x^2 - v_{0x}^2}{2(a_x)_{\text{max}}} = \frac{0 - v_{0x}^2}{-2\mu_s g} = \frac{v_{0x}^2}{2\mu_s g} = \frac{(12.0 \text{ m/s})^2}{2(0.500)(9.80 \text{ m/s}^2)} = 14.7 \text{ m}$$ ◊

(b) Observe that neither the mass of the truck nor the mass of the load were needed in the solution of this problem. ◊

49. An object falling under the pull of gravity is acted upon by a frictional force of air resistance. The magnitude of this force is approximately proportional to the speed of the object, which can be written as $f = bv$. Assume that $b = 15$ kg/s and $m = 50$ kg. (a) What is the terminal speed the object reaches while falling? (b) Does your answer to part (a) depend on the initial speed of the object? Explain.

Solution

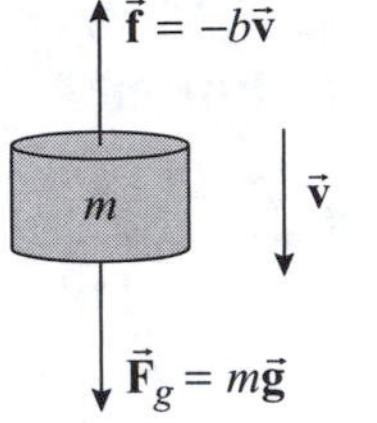

(a) A force diagram of the falling body is given at the right. Taking downward as positive, Newton's second law yields

$$\Sigma F_y = mg - bv = ma_y$$

or the downward acceleration of the falling body will be

$$a_y = \frac{mg - bv}{m} = g - \left(\frac{b}{m}\right)v$$

When the falling body achieves a constant downward speed (the terminal speed), the acceleration is $a_y = 0$, or

$$g - \left(\frac{b}{m}\right)v_{\text{terminal}} = 0$$

and the terminal speed is

$$v_{\text{terminal}} = \frac{g}{(b/m)} = \frac{mg}{b} = \frac{(50\text{ kg})(9.80\text{ m/s}^2)}{15\text{ kg/s}} = 33\text{ m/s} \quad \Diamond$$

(b) The magnitude of the terminal speed does not depend on the initial speed of the falling body. However, the manner in which the body approaches its terminal speed does depend on the initial speed. If $v_0 < v_{\text{terminal}}$ (as when a body starts falling from rest), then $f < mg$ and the object has a downward acceleration, picking up speed until it reaches a maximum value of $v = v_{\text{terminal}}$. On the other hand, if $v_0 > v_{\text{terminal}}$ (as might be the case of a skydiver just after opening the parachute), then $f > mg$ and the object has an upward acceleration, slowing until it reaches a minimum value of $v = v_{\text{terminal}}$. Finally, if $v_0 = v_{\text{terminal}}$, then $f = mg$ and the object has zero acceleration, meaning that it continues to fall at a constant speed of $v_0 = v_{\text{terminal}}$. ◊

53. To meet a U.S. Postal Service requirement, employees' footwear must have a coefficient of static friction of 0.500 or more on a specified tile surface. A typical athletic shoe has a coefficient of 0.800. In an emergency, what is the minimum time interval in which a person starting from rest can move 3.00 m on the tile surface if she is wearing (a) footwear meeting the Postal Service minimum and (b) a typical athletic shoe?

Solution

When a person walks or runs across a surface, they push backward on that surface through a friction force exerted on the surface by the sole of their shoe. In accordance with Newton's third law, the surface then exerts a reaction force in the forward direction on the person. The magnitude of this reaction force is the same as the magnitude of the friction force.

Starting from rest ($v_0 = 0$), the time required to move distance Δx across a floor is given by $\Delta x = v_0 t + \frac{1}{2}at^2 = 0 + \frac{1}{2}at^2$ as $t = \sqrt{2\Delta x/a}$. Thus, for a fixed distance, such as $\Delta x = 3.00$ m, the movement is made in minimum time when the maximum possible acceleration is maintained. The maximum possible acceleration without slipping is given by

$$a_{max} = \frac{(f_s)_{max}}{m} = \frac{\mu_s n}{m} = \frac{\mu_s (mg)}{m} = \mu_s g$$

where f_s is the static friction force between the shoe sole and the floor, $n = mg$ is the normal force exerted on the person by the floor, and μ_s is the coefficient of static friction. The minimum time required to move distance Δx without slipping is then

$$t_{min} = \sqrt{\frac{2\Delta x}{a_{max}}} = \sqrt{\frac{2\Delta x}{\mu_s g}}$$

(a) If $\Delta x = 3.00$ m and $\mu_s = 0.500$, the minimum time is

$$t_{min} = \sqrt{\frac{2(3.00\text{ m})}{(0.500)(9.80\text{ m/s}^2)}} = 1.11\text{ s} \qquad \lozenge$$

(b) When $\mu_s = 0.800$, the minimum time to move $\Delta x = 3.00$ m is

$$t_{min} = \sqrt{\frac{2(3.00\text{ m})}{(0.800)(9.80\text{ m/s}^2)}} = 0.875\text{ s} \qquad \lozenge$$

58. The force exerted by the wind on a sailboat is approximately perpendicular to the sail and proportional to the component of the wind velocity perpendicular to the sail. For the 800-kg sailboat shown in Figure P4.58, the force exerted by the wind on the sailboat is

$$F_{\text{sail}} = \left(550\ \frac{\text{N}}{\text{m/s}}\right) v_{\text{wind}\perp}$$

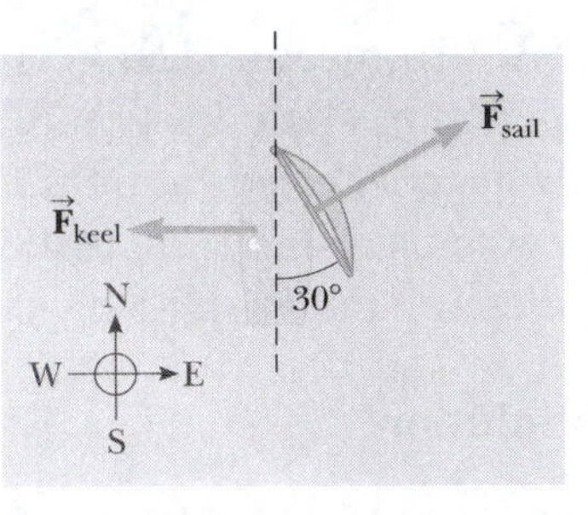

Figure P4.58

Water exerts a force along the keel (bottom) of the boat that prevents it from moving sideways, as shown in the figure. Once the boat starts moving forward, water also exerts a drag force backwards on the boat, opposing the forward motion. If a 17-knot wind (1 knot = 0.514 m/s) is blowing to the east, what is the initial acceleration of the sailboat?

Solution

The heavy line in the sketch at the right represents the edge of the sail as seen from above. The wind blows toward the east with velocity $\vec{\mathbf{v}}_{\text{wind}}$, and exerts a force $\vec{\mathbf{F}}_{\text{sail}}$ perpendicular to the sail. The magnitude of this force is

$$F_{\text{sail}} = \left(550\ \frac{\text{N}}{\text{m/s}}\right) v_{\text{wind}\perp}$$

where $v_{\text{wind}\perp}$ is the component of the wind velocity perpendicular to the sail. When the sail is oriented at 30° from the north-south line and the wind speed is $v_{\text{wind}} = 17$ knots,

$$F_{\text{sail}} = \left(550\ \frac{\text{N}}{\text{m/s}}\right) v_{\text{wind}\perp} = \left(550\ \frac{\text{N}}{\text{m/s}}\right)\left[\left(17\ \text{knots}\right)\left(\frac{0.514\ \text{m/s}}{1\ \text{knot}}\right)\cos 30°\right] = 4.2 \times 10^3\ \text{N}$$

A westward force $\vec{\mathbf{F}}_{\text{keel}}$ (see Figure P4.58) acts on the bottom of the boat to keep it from moving sideways by counterbalancing the eastward component of $\vec{\mathbf{F}}_{\text{sail}}$. Initially, before the boat has gained enough speed to develop any significant drag force, the acceleration is provided by the northward component of $\vec{\mathbf{F}}_{\text{sail}}$. This initial acceleration is

$$a = \frac{\left|\vec{\mathbf{F}}_{\text{sail}}\right|_{\text{north}}}{m} = \frac{\left(4.2 \times 10^3\ \text{N}\right)\sin 30°}{800\ \text{kg}} = \boxed{2.6\ \text{m/s}^2}$$

◊

67. A 2.00-kg aluminum block and a 6.00-kg copper block are connected by a light string over a frictionless pulley. The two blocks are allowed to move on a fixed steel block wedge (of angle $\theta = 30.0°$) as shown in Figure P4.67. Making use of Table 4.2, determine (a) the acceleration of the two blocks and (b) the tension in the string.

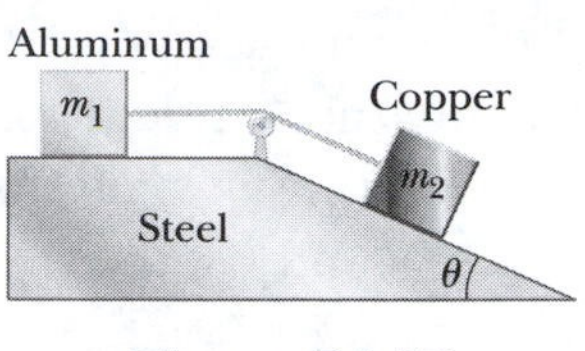

Figure P4.67

Solution

Force diagrams of each block are given below. In each diagram, the axes have been chosen to be parallel to and perpendicular to the surface across which the block moves.

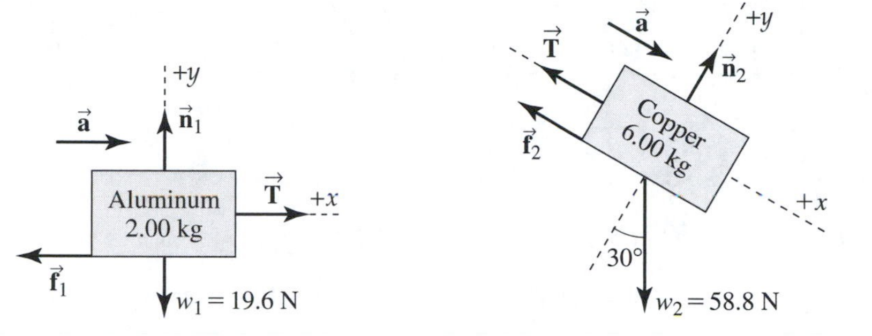

From Table 4.2, we find the coefficient of kinetic friction for aluminum on steel to be $\mu_{k1} = 0.47$ and that for copper on steel to be $\mu_{k2} = 0.36$. Since $a_y = 0$ for each block, the normal forces are $n_1 = w_1$ and $n_2 = w_2 \cos 30.0°$. The magnitudes of the kinetic friction forces are then

$f_1 = \mu_{k1} n_1 = (0.47)(19.6\text{ N}) = 9.2\text{ N}$ and $f_2 = \mu_{k2} n_2 = (0.36)(58.8\text{ N})\cos 30.0° = 18\text{ N}$

Applying Newton's second law to each block yields the following equations:

For Aluminum block: $\Sigma F_x = ma_x \Rightarrow T - f_1 = m_1 a$ or $T = 9.2\text{ N} + (2.00\text{ kg})a$ **[1]**

For Copper block: $\Sigma F_x = ma_x \Rightarrow (58.8\text{ N})\sin 30.0° - f_2 - T = m_2 a$, or

$29.4\text{ N} - 18\text{ N} - T = (6.00\text{ kg})a$ which reduces to $11\text{ N} - T = (6.00\text{ kg})a$ **[2]**

(a) Substituting Equation [1] into [2] and solving for the acceleration gives

$$a = \frac{11\text{ N} - 9.2\text{ N}}{8.00\text{ kg}} = 0.23\ \text{m/s}^2$$ ◊

(b) Substituting this result into Equation [1] yields $T = 9.7$ N. ◊

73. A van accelerates down a hill (Fig. P4.73), going from rest to 30.0 m/s in 6.00 s. During the acceleration, a toy $(m = 0.100\text{ kg})$ hangs by a string from the van's ceiling. The acceleration is such that the string remains perpendicular to the ceiling. Determine (a) the angle θ and (b) the tension in the string.

Solution

(a) The sketch at the right gives a force diagram of the van, with the positive x-direction being down the incline. Since the van starts from rest and reaches a velocity of 30.0 m/s down the incline in 6.00 s, its acceleration is given by $v_x = v_{0x} + a_x t$ as

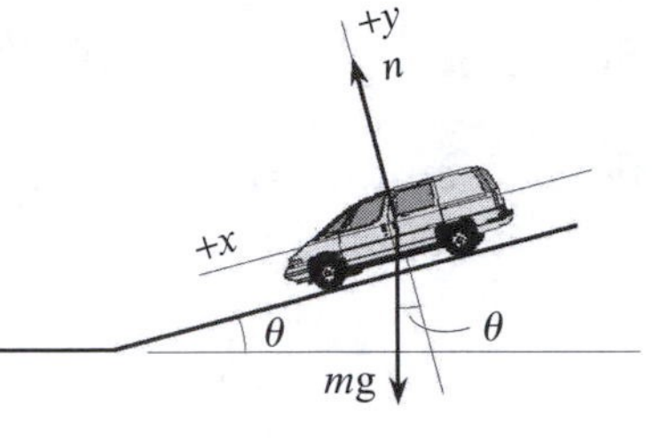

$$a_x = \frac{v_x - v_{0x}}{t} = \frac{30.0\text{ m/s} - 0}{6.00\text{ s}} = 5.00\text{ m/s}^2$$

Applying Newton's second law for the x-direction gives

$$\Sigma F_x = mg\sin\theta = ma_x \quad \text{or} \quad a_x = g\sin\theta$$

Therefore, the angle of the incline must be

$$\theta = \sin^{-1}\left(\frac{a_x}{g}\right) = \sin^{-1}\left(\frac{5.00\text{ m/s}^2}{9.80\text{ m/s}^2}\right) = 30.7°$$ ◊

(b) Now consider a force diagram of the suspended toy. Since the string remains perpendicular to the ceiling of the van, it must lie along the y-axis in the sketch above, or it is at angle θ from the vertical as seen in the sketch at the right.

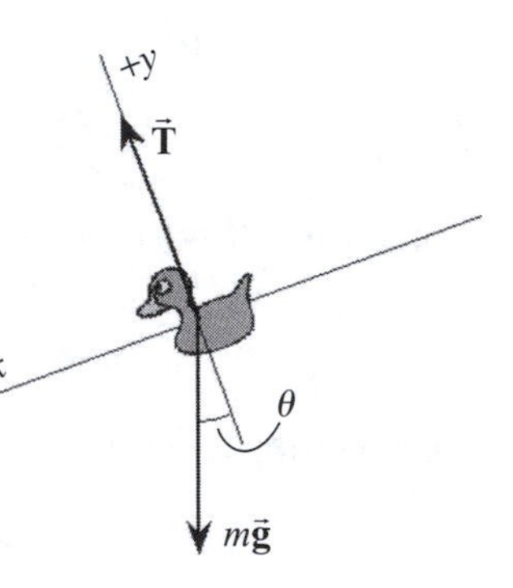

The acceleration of the van and its contents is directed down the incline, or in the positive x-direction. Thus, $a_y = 0$ and Newton's second law gives

$$\Sigma F_y = T - mg\cos\theta = 0$$

or

$$T = mg\cos\theta = (0.100\text{ kg})(9.80\text{ m/s}^2)\cos 30.7° = 0.843\text{ N}$$ ◊

78. A sled weighing 60.0 N is pulled horizontally across snow so that the coefficient of kinetic friction between sled and snow is 0.100. A penguin weighing 70.0 N rides on the sled, as in Figure P4.78. If the coefficient of static friction between penguin and sled is 0.700, find the maximum horizontal force that can be exerted on the sled before the penguin begins to slide off.

Figure P4.78

Solution

Two force diagrams, one for the penguin alone and one for the penguin and sled together, are given below:

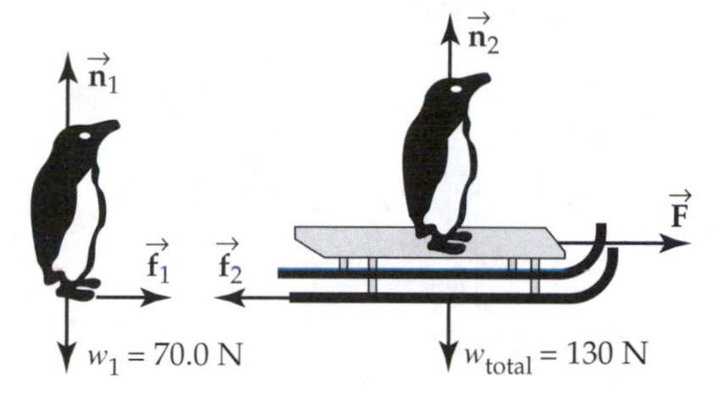

The force which will accelerate the penguin forward, without slipping, is the static friction force $\vec{\mathbf{f}}_1$ between his feet and the sled. The maximum magnitude of this force is $f_{1,\max} = \mu_s n_1 = \mu_s w_1$. Thus, the maximum acceleration the penguin can be given before he begins to slide off the sled is

$$a_{\max} = \frac{f_{1,\max}}{m_1} = \frac{\mu_s \cancel{m_1} g}{\cancel{m_1}} = \mu_s g$$

Until the penguin starts to slip on the sled, the penguin and the sled have the same acceleration. Applying Newton's second law to the combined system of sled plus penguin gives

$$\Sigma F_x = F - f_2 = m_{\text{total}} a \qquad \text{or} \qquad F = f_2 + m_{\text{total}} a$$

where $n_2 = w_2 = m_{\text{total}} g$ is the normal force acting on this system, while $f_2 = \mu_k n_2 = \mu_k m_{\text{total}} g$ is the kinetic friction force between the sled runners and the snow. The magnitude of the applied force $\vec{\mathbf{F}}$, when $a = a_{\max} = \mu_s g$ and the penguin is on the verge of slipping, is given by

$$F = \mu_k m_{\text{total}} g + m_{\text{total}} a_{\max} = \mu_k m_{\text{total}} g + m_{\text{total}} \mu_s g = (\mu_k + \mu_s) m_{\text{total}} g$$

or

$$F = (\mu_k + \mu_s) w_{\text{total}} = (0.100 + 0.700)(130 \text{ N}) = 104 \text{ N} \qquad \Diamond$$

5

Energy

NOTES FROM SELECTED CHAPTER SECTIONS

5.1 Work

In order for work to be accomplished, an object must undergo a displacement; the force associated with the work must have a component parallel to the direction of the displacement. Work is a scalar quantity and can be either positive or negative (positive when the component $F\cos\theta$ is in the same direction as the displacement). The SI unit of work is the newton-meter ($N \cdot m$) or joule (J).

5.2 Kinetic Energy and the Work-Energy Theorem

Any object that has mass m and speed v has **kinetic energy**. Kinetic energy is a scalar quantity and has the same units as work; kinetic energy of an object will change only if net work is done on the object by external forces. The relationship between work and change in kinetic energy is stated in the work-energy theorem.

A force is **conservative** if the work it does on an object moving between two points is independent of the path the object takes between the points. The work done on an object by a conservative force depends only on the initial and final positions of the object. The gravitational force is an example of a conservative force.

A force is **nonconservative** if the work it does on an object moving between two points depends on the path taken. Kinetic friction is an example of a nonconservative force.

5.3 Gravitational Potential Energy

The work done on an object by the force of gravity is equal to the object's initial potential energy minus its final potential energy ($W_g = -\Delta PE_g$). The gravitational potential energy associated with an object depends only on the object's weight and its vertical height above the surface of the Earth. If the height above the surface increases, the potential energy will also increase, but the work done by the gravitational force will be negative. (In this case the direction of the displacement is opposite the direction of the gravitational force.) **In working problems involving gravitational potential energy, it is necessary to choose an arbitrary reference level (or location) at which the potential energy is taken to be zero.**

5.4 Spring Potential Energy

Elastic potential energy is the energy associated with a spring that is compressed from or stretched beyond its equilibrium position. The spring force is a conservative force and has a magnitude that is proportional to the displacement of the spring from the equilibrium position.

5.5 Systems and Energy Conservation

The **change in the kinetic energy** of a physical system equals the sum of the work done by conservative forces and the work done by nonconservative forces. The sum of the kinetic energy plus the potential energy is called the **total mechanical energy**. Since the work done by conservative forces equals the negative of the change in potential energy ($W_c = -\Delta PE$), the work done by nonconservative forces equals the change in the total mechanical energy of the system. *The mechanical energy of a system remains constant if only conservative forces do work on the system.*

5.6 Power

Power delivered to an object is defined as the rate at which energy is transferred to the object or the rate at which work is being done on the object. The average power delivered to an object during a time interval can be expressed as the product of the average speed during the time interval and the component of the force in the direction of the velocity.

5.7 Work Done by a Varying Force

If the value of a non-constant force is known as a function of position, the work done by the varying force during a displacement can be determined by calculating the area under the force vs. displacement curve.

EQUATIONS AND CONCEPTS

The **work done by a constant force $\vec{\mathbf{F}}$**, (constant in both magnitude and direction) is defined to be the product of the component of the force in the direction of the displacement and the magnitude of the displacement.

$$W = (F\cos\theta)\Delta x \quad (5.3)$$

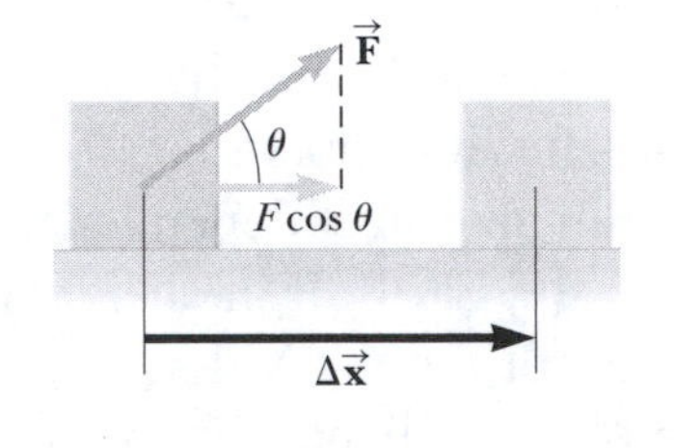

Work is a scalar quantity and the SI unit of work is the newton-meter or joule.

$$1\ \text{N}\cdot\text{m} = 1\ \text{J}$$

$$\text{joule (J)} = \text{newton}\cdot\text{meter} = \text{kg}\cdot\text{m}^2/\text{s}^2$$

Work can be positive, negative, or zero, depending on the value of θ, the angle between the direction of the force and the direction of the displacement. The figure at the right shows an object acted on by four forces (gravitational force, applied force, normal force, and spring force) while being displaced down a smooth incline. Consider the work done by each of the four forces as illustrated below *when the direction of the x-axis is taken to be along the direction of the incline.*

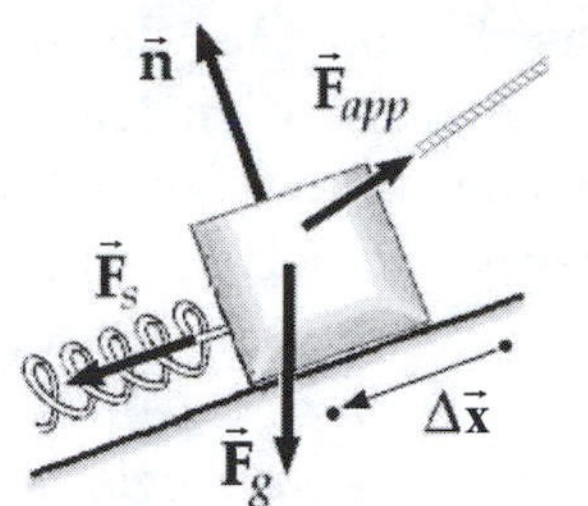

$\Delta\vec{\mathbf{x}}$ $\vec{\mathbf{F}}_g$

$0 \le \theta < 90°$
W_g is positive

$90° \le \theta < 180°$
W_{app} is negative

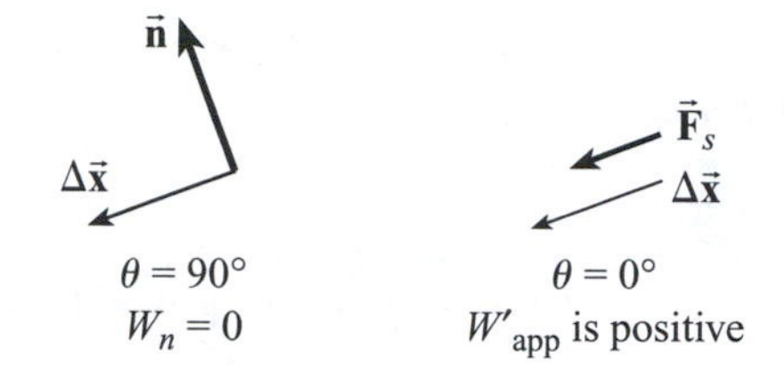

$\theta = 90°$
$W_n = 0$

$\theta = 0°$
W'_{app} is positive

The **work-energy theorem** states that the net work done on a particle equals the change in kinetic energy of the particle. Kinetic energy is defined by Equation 5.6. It is a scalar quantity associated with the motion of a mass and has the same units as work. *The work-energy theorem is valid for a particle or for a system that can be modeled as a particle.*

$$W_{net} = \tfrac{1}{2}mv^2 - \tfrac{1}{2}mv_0^2 \quad (5.5)$$

$$KE \equiv \tfrac{1}{2}mv^2 \quad (5.6)$$

$$W_{net} = KE_f - KE_i = \Delta KE \quad (5.7)$$

The **net work** done on an object is the sum of the work done by the nonconservative forces and the conservative forces.

$$W_{nc} + W_c = \Delta KE \quad (5.8)$$

Gravitational potential energy of a mass-Earth system near the surface of the Earth is defined by Equation 5.11. *The quantity y in Equation 5.11 is the vertical position of the mass relative to the Earth (or other arbitrarily chosen reference level where $PE_g = 0$).*

$$PE \equiv mgy \quad (5.11)$$

The **work done by the force of gravity** can be expressed in terms of initial and final values of the y-coordinates. *The units of energy (kinetic and potential) are the same as the units of work. The difference in potential energy between two points is independent of the location of the origin.*

$$W_g = -\left(PE_f - PE_i\right) = -\left(mgy_f - mgy_i\right) \quad (5.12)$$

Hooke's law expresses the force exerted by a spring that is stretched or compressed from the equilibrium position. *The negative sign signifies that the force is always directed opposite the displacement from equilibrium.*

$$F_s = -kx \quad (5.16)$$

k = spring constant in N/m.

Each spring has a characteristic value of k.

Elastic potential energy is associated with a mass-spring system that has been stretched or compressed from the equilibrium position. *The elastic potential energy of a deformed spring is always positive.*

$$PE_s \equiv \tfrac{1}{2}kx^2 \quad (5.17)$$

x = displacement from equilibrium.

The **work done by all nonconservative forces** equals the change in the total mechanical energy of a system (change in kinetic energy plus changes in the gravitational and elastic potential energies).

$$W_{nc} = (KE_f - KE_i) + (PE_{gf} - PE_{gi}) + (PE_{sf} - PE_{si}) \quad (5.18)$$

$$W_{nc} = \Delta KE + \Delta PE_g + \Delta PE_s = \Delta E_{total}$$

The law of conservation of mechanical energy holds when *only conservative* forces act on a system; the total mechanical energy of the system remains constant.

$$(KE + PE_g + PE_s)_i = (KE + PE_g + PE_s)_f \quad (5.19)$$

The **average power** supplied by a force is the ratio of the work done by the force to the time interval over which the force acts. The average power can also be expressed in terms of the force and the average speed of the object on which the force acts. *In Equation 5.23, F is the component of the force along the direction of the velocity.*

$$\bar{P} \equiv \frac{W}{\Delta t} \quad (5.23)$$

$$\bar{P} = F\bar{v} \quad (5.24)$$

The **SI unit of power** is the watt; in the U.S. customary system, the unit of power is the horsepower.

$$1\ \text{W} = 1\ \text{J/s} = 1\ \text{kg}\cdot\text{m}^2/\text{s}^3 \quad (5.26a)$$

$$1\ \text{hp} \equiv 550\ \frac{\text{ft}\cdot\text{lb}}{\text{s}} = 746\ \text{W} \quad (5.26b)$$

Work done by a variable force is equal to the area under the force vs. displacement curve. The figure at right illustrates the case of a force acting along the x-axis.

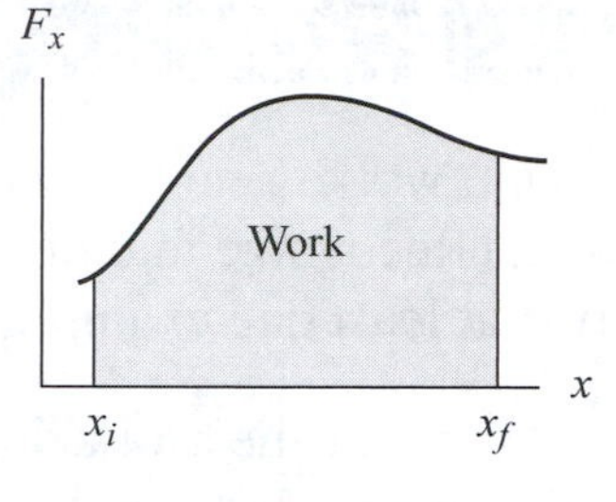

SUGGESTIONS, SKILLS, AND STRATEGIES

Choosing a zero level for potential energy

In working problems involving gravitational potential energy, it is always necessary to choose a location at which to set the gravitational potential energy equal to zero. This choice is completely arbitrary because the important quantity is the difference in potential energy, and that difference is independent of the location of zero. It is often convenient, but not essential, to choose the surface of the Earth as the reference position for zero potential energy. In most cases, the statement of the problem suggests a convenient level to use.

Conservation of energy

Take the following steps in applying the principle of conservation of energy:

1. Define your system, which may consist of more than one object.

2. Select a reference level for the zero point of gravitational potential energy. This level must not be changed during the solution of a specific problem.

3. Determine whether or not nonconservative forces are present.

4. If mechanical energy is conserved (i.e., if only conservative forces are present), you should use Equation 5.18, $(KE + PE_g + PE_s)_i = (KE + PE_g + PE_s)_f$. Usually (except when a quantity is equal to zero) it is better to solve for the unknown algebraically (using symbols) before substituting numerical values.

5. If nonconservative forces such as friction are present (and thus mechanical energy is not conserved), first write expressions for the total initial and total final mechanical energies. In this case, the change in the two total energies is equal to the work done by the nonconservative force(s).

REVIEW CHECKLIST

- Define the work done by a constant force and work done by a force that varies with position. (Recognize that the work done by a force can be positive, negative, or zero; describe at least one example of each case.)

- Understand that the work done by a conservative force in moving a body between any two points is independent of the path taken. Nonconservative forces are those for which the work done on a particle moving between two points depends on the path. Account for nonconservative forces acting on a system using the work-energy theorem. In this case, the work done by all nonconservative forces equals the change in total mechanical energy of the system.

- Relate the work done by the net force on an object to the **change in kinetic energy**. The relation $W_{\text{net}} = \Delta KE = KE_f - KE_i$ is called the work-energy theorem, and it is valid whether or not the (resultant) force is constant. That is, if we know the net work done on a particle as it undergoes a displacement, we also know the change in its kinetic energy. This is the most important concept in this chapter, so you must understand it thoroughly.

- Recognize that the gravitational potential energy of the mass-Earth system, $PE_g = mgy$, can be positive, negative, or zero, depending on the location of the reference level used to measure y. Be aware of the fact that although PE depends on the origin of the coordinate system, *the change in potential energy,* $(PE)_f - (PE)_i$, *is independent of the coordinate system used to define PE.*

- Calculate average power when work is accomplished over a time interval and instantaneous power when an applied force acts on an object moving with speed v.

SOLUTIONS TO SELECTED END-OF-CHAPTER PROBLEMS

7. A sledge loaded with bricks has a total mass of 18.0 kg and is pulled at constant speed by a rope inclined at 20.0° above the horizontal. The sledge moves a distance of 20.0 m on a horizontal surface. The coefficient of kinetic friction between the sledge and surface is 0.500. (a) What is the tension in the rope? (b) How much work is done by the rope on the sledge? (c) What is the mechanical energy lost due to friction?

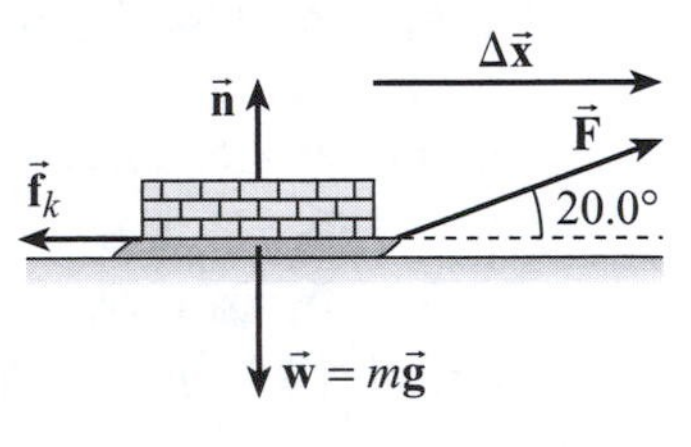

Solution

(a) The sledge has constant velocity, so

$$\Sigma F_y = n + F\sin 20.0° - mg = ma_y = 0 \quad \Rightarrow \quad n = mg - F\sin 20.0°$$

Thus, the kinetic friction force is

$$f_k = \mu_k n = \mu_k\left(mg - F\sin 20.0°\right)$$

Also,

$$\Sigma F_x = F\cos 20.0° - f_k = ma_x = 0, \text{ or}$$
$$F\cos 20.0° - \mu_k(mg - F\sin 20.0°) = 0$$

Simplifying this result gives the tension in the rope as

$$F = \frac{\mu_k mg}{\cos 20.0° + \mu_k \sin 20.0°} = \frac{0.500(18.0\text{ kg})(9.80\text{ m/s}^2)}{\cos 20.0° + 0.500\sin 20.0°} = 79.4\text{ N}$$ ◊

(b) The work done by the rope on the sledge is $W_F = (F\cos 20.0°)\Delta x$, or

$$W_F = (79.4\text{ N})(20.0\text{ m})\cos 20.0° = 1.49\times 10^3\text{ J} = 1.49\text{ kJ}$$ ◊

(c) The friction force and the displacement are in opposite directions ($\theta = 180°$), so the work done by friction on the sledge is

$$W_{f_k} = (f_k\cos 180°)\Delta x = \mu_k(mg - F\sin 20.0°)(\Delta x)\cos 180°$$

$$= 0.500\left[(18.0\text{ kg})(9.80\text{ m/s}^2) - (79.4\text{ N})\sin 20.0°\right](20.0\text{ m})(-1)$$

or

$$W_{f_k} = -1.49\times 10^3\text{ J} = -1.49\text{ kJ}$$ ◊

---

15. A 7.80-g bullet moving at 575 m/s penetrates a tree trunk to a depth of 5.50 cm. (a) Use work and energy considerations to find the average frictional force that stops the bullet. (b) Assuming the frictional force is constant, determine how much time elapses between the moment the bullet enters the tree and the moment it stops moving.

Solution

(a) The frictional (or resistance) force is the only force that does work on the bullet as the bullet penetrates the tree. This force is directed opposite to the motion of the bullet, so the work done by the average frictional force is $W_{\text{friction}} = (f_{\text{av}}\cos\theta)\Delta x$, where $\theta = 180°$.

From the work-energy theorem, $W_{\text{net}} = KE_f - KE_i$, with $W_{\text{net}} = W_{\text{friction}}$, we find

$$(f_{\text{av}}\cos\theta)\Delta x = KE_f - KE_i = \frac{1}{2}mv_f^2 - \frac{1}{2}mv_i^2$$

Since the bullet stops, $v_f = 0$, and we are given that $v_i = 575\text{ m/s}$, while the mass of the bullet is $m = 7.80\text{ g} = 7.80\times 10^{-3}\text{ kg}$. If the bullet is brought to rest in a distance $\Delta x = 5.50\text{ cm} = 5.50\times 10^{-2}\text{ m}$, the average frictional force must be

$$f_{\text{av}} = \frac{KE_f - KE_i}{(\Delta x)\cos\theta} = \frac{0 - \frac{1}{2}(7.80\times 10^{-3}\text{ kg})(575\text{ m/s})^2}{(5.50\times 10^{-2}\text{ m})\cos 180°} = 2.34\times 10^4\text{ N}$$ ◊

(b) If the frictional force may be assumed to be constant, then the bullet has a constant acceleration. We may then use any of the kinematics equations from Chapter 2 to solve for the time required to stop the bullet. Since we know both the initial and final velocities of the bullet, as well as the displacement of the bullet during the time of interest, the easiest of the kinematics equations to use is $\Delta x = v_{av}\Delta t$, where $v_{av} = (v_i + v_f)/2$. From this relation, we find

$$\Delta t = \frac{\Delta x}{v_{av}} = \frac{2\Delta x}{v_i + v_f} = \frac{2(5.50\times 10^{-2}\ \text{m})}{575\ \text{m/s} + 0} = 1.91\times 10^{-4}\ \text{s} = 0.191\ \text{ms}$$ ◊

---

20. When a 2.50-kg object is hung vertically on a certain light spring described by Hooke's law, the spring stretches 2.76 cm. (a) What is the force constant of the spring? (b) If the 2.50-kg object is removed, how far will the spring stretch if a 1.25-kg block is hung on it? (c) How much work must an external agent do to stretch the same spring 8.00 cm from its unstretched position?

Solution

(a) When an object hangs in equilibrium on the end of a spring, the upward spring force acting on the object must counterbalance the downward gravitational force, or weight. From Hooke's law, the magnitude of the spring force is $F_s = kx$, where k is the force constant of the spring and x is the distance the spring is deformed (either stretched or compressed) from its natural length. The magnitude of the gravitational force is $F_g = mg$, so requiring that $F_s = F_g$ yields $kx = mg$, or

$$k = \frac{mg}{x} = \frac{(2.50\ \text{kg})(9.80\ \text{m/s}^2)}{2.76\times 10^{-2}\ \text{m}} = 888\ \text{N/m}$$ ◊

(b) When the mass of the object suspended on the end of the spring is $m = 1.25$ kg, the elongation of the spring will be

$$x = \frac{mg}{k} = \frac{(1.25\ \text{kg})(9.80\ \text{m/s}^2)}{888\ \text{N/m}} = 1.38\times 10^{-2}\ \text{m} = 1.38\ \text{cm}$$ ◊

(c) The work required to stretch a spring distance x from its unstretched state is equal to the elastic potential energy ($PE_s = \frac{1}{2}kx^2$) stored in the spring in its stretched state. Thus, if the spring described above is to be stretched by 8.00 cm, the work required will be

$$W = PE_s = \frac{1}{2}kx^2 = \frac{1}{2}(888\ \text{N/m})(8.00\times 10^{-2}\ \text{m})^2 = 2.84\ \text{N}\cdot\text{m} = 2.84\ \text{J}$$ ◊

---

29. A 50.0-kg projectile is fired at an angle of 30.0° above the horizontal with an initial speed of 1.20×10^2 m/s from the top of a cliff 142 m above level ground, where the ground is taken to be $y = 0$. (a) What is the initial total mechanical energy of the projectile? (b) Suppose the projectile is traveling 85.0 m/s at its maximum height of $y = 427$ m. How much work has been done on the projectile by air friction? (c) What is the speed of the projectile immediately before it hits the ground if air friction does one and a half times as much work on the projectile when it is going down as it did when it was going up?

Solution

(a) Taking $y = 0$, and hence $PE_g = 0$, at ground level, the mechanical energy of the projectile as it leaves the launch point is

$$E_i = KE_i + (PE_g)_i = \frac{1}{2}mv_i^2 + mgy_i$$

$$= \frac{1}{2}(50.0 \text{ kg})(1.20 \times 10^2 \text{ m/s})^2 + (50.0 \text{ kg})(9.80 \text{ m/s}^2)(142 \text{ m}) = 4.30 \times 10^5 \text{ J} \quad \Diamond$$

(b) If the projectile has a speed of $v_{\text{peak}} = 85.0$ m/s when it reaches its maximum altitude at $y_{\text{peak}} = 427$ m above ground level, the work done on the projectile by nonconservative forces as it was rising has been

$$(W_{nc})_{\text{rise}} = \Delta KE + \Delta PE_g = \frac{1}{2}m\left(v_{\text{peak}}^2 - v_i^2\right) + mg\left(y_{\text{peak}} - y_i\right)$$

or

$$(W_{nc})_{\text{rise}} = \frac{1}{2}(50.0 \text{ kg})\left[(85.0 \text{ m/s})^2 - (1.20 \times 10^2)^2\right]$$
$$+ (50.0 \text{ kg})(9.80 \text{ m/s}^2)(427 \text{ m} - 142 \text{ m})$$

which yields

$$(W_{nc})_{\text{rise}} = -3.97 \times 10^4 \text{ J} \quad \Diamond$$

(c) If air friction does one and a half times as much work on the projectile while the projectile falls to the ground as it did while the projectile was rising, the total work done on the projectile by air friction during its flight is

$$(W_{nc})_{\text{total}} = (W_{nc})_{\text{rise}} + (W_{nc})_{\text{fall}} = (W_{nc})_{\text{rise}} + 1.50(W_{nc})_{\text{rise}}$$

$$= 2.50(W_{nc})_{\text{rise}} = 2.50(-3.97 \times 10^4 \text{ J}) = -9.93 \times 10^4 \text{ J}$$

Applying the work-energy theorem, $(W_{nc})_{\text{total}} = (KE_f - KE_i) + (PE_g)_f - (PE_g)_i$, to the full flight of the projectile, we obtain $KE_f = KE_i + (PE_g)_i - (PE_g)_f + (W_{nc})_{\text{total}}$

or

$$\frac{1}{2}mv_f^2 = \frac{1}{2}mv_i^2 + mg(y_i - y_f) + (W_{nc})_{\text{total}}$$

The speed of the projectile just before it hits the ground is then

$$v_f = \sqrt{v_i^2 + 2g\left(y_i - y_f\right) + \frac{2\left(W_{nc}\right)_{\text{total}}}{m}}$$

$$= \sqrt{\left(1.20\times10^2\ \text{m/s}\right)^2 + 2\left(9.80\ \text{m/s}^2\right)(142\ \text{m} - 0) + \frac{2\left(-9.93\times10^4\ \text{J}\right)}{50.0\ \text{kg}}}$$

or

$$v_f = 115\ \text{m/s}$$ ◊

---

37. Tarzan swings on a 30.0-m-long vine initially inclined at an angle of 37.0° with the vertical. What is his speed at the bottom of the swing (a) if he starts from rest? (b) if he pushes off with a speed of 4.00 m/s?

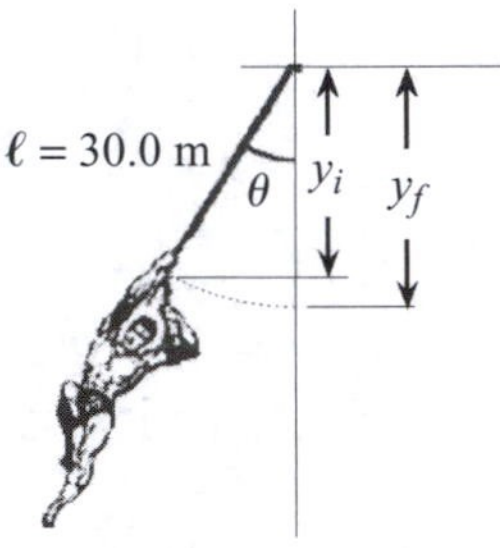

Solution

If we ignore air resistance, the only nonconservative force acting on Tarzan, as he swings from the initial position shown to the bottom of his arc, is the tension in the vine. However, this tension is directed toward the center of his circular path, and at each point on his path, is perpendicular to his motion. This means that the tension force does no work on Tarzan, or $W_{nc} = 0$.

The work-energy theorem then takes the form

$$W_{nc} = \left(KE_f - KE_i\right) + \left(PE_f - PE_i\right) = 0 \quad \text{or} \quad KE_f = KE_i + \left(PE_i - PE_f\right)$$

Since $KE = \frac{1}{2}mv^2$, and the potential energy is gravitational, $PE = mgy$, this becomes

$$\frac{1}{2}\cancel{m}v_f^2 = \frac{1}{2}\cancel{m}v_i^2 + \cancel{m}g\left(y_i - y_f\right)$$

Realizing that Tarzan's y coordinate is $y = -\ell\cos\theta = -(30.0\ \text{m})\cos\theta$, we see that

$$y_i - y_f = -(30.0\ \text{m})\left(\cos\theta_i - \cos\theta_f\right) = -(30.0\ \text{m})(\cos 37.0° - \cos 0°) = +6.04\ \text{m}$$

and his speed at the lowest point on the swing is

$$v_f = \sqrt{v_i^2 + 2g(6.04\ \text{m})}$$

(a) If he starts from rest $\left(v_i = 0\right)$,

$$v_f = \sqrt{0 + 2\left(9.80\ \text{m/s}^2\right)(6.04\ \text{m})} = 10.9\ \text{m/s}$$ ◊

(b) If $v_i = 4.00$ m/s,

$$v_f = \sqrt{(4.00\ \text{m/s})^2 + 2\left(9.80\ \text{m/s}^2\right)(6.04\ \text{m})} = 11.6\ \text{m/s}$$ ◊

---

45. A 2.1×10^3-kg car starts from rest at the top of a 5.0-m-long long driveway that is inclined at 20° with the horizontal. If an average friction force of 4.0×10^3 N impedes the motion, find the speed of the car at the bottom of the driveway.

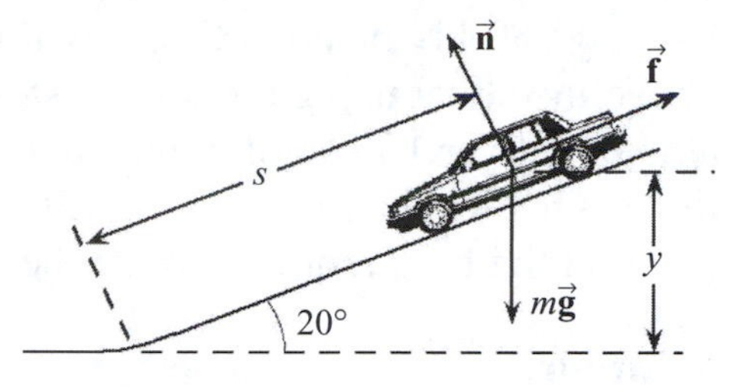

Solution

We shall choose our zero gravitational potential energy level ($y = 0$) to be at the level of the bottom of the driveway. Then, the initial height of the car above this reference level is $y_i = s \cdot \sin\theta$. As the car rolls down the incline, the normal force is always perpendicular to the motion and hence does no work. The gravitational force, $m\vec{\mathbf{g}}$, is a conservative force, so the only nonconservative force doing work on the car is the frictional force $\vec{\mathbf{f}}$. The work done by this force as the car rolls down the driveway is

$$W_{nc} = (f\cos\theta) \cdot s = (f\cos 180°) \cdot s = -f \cdot s$$

Since the car starts from rest, $KE_i = \frac{1}{2}mv_i^2 = 0$, and when the car reaches the bottom of the driveway, $PE_{g,f} = mgy_f = mg(0) = 0$. The work-energy theorem

$$W_{nc} = (KE_f + PE_f) - (KE_i + PE_i) = \left(\tfrac{1}{2}mv_f^2 + mgy_f\right) - \left(\tfrac{1}{2}mv_i^2 + mgy_i\right)$$

then gives

$$-f \cdot s = \left(\tfrac{1}{2}mv_f^2 + 0\right) - (0 + mgy_i) = \tfrac{1}{2}mv_f^2 - mgs \cdot \sin\theta$$

Solving for the speed of the car at the bottom of the driveway, we have

$$v_f = \sqrt{2s(g\sin\theta - f/m)}$$

Thus, if $s = 5.0$ m, $\theta = 20°$, $m = 2.1 \times 10^3$ kg, and $f = 4.0 \times 10^3$ N, the final speed of the car is

$$v_f = \sqrt{2(5.0\text{ m})\left[(9.80\text{ m/s}^2)\sin 20° - \frac{4.0 \times 10^3\text{ N}}{2.1 \times 10^3\text{ kg}}\right]} = 3.8\text{ m/s}$$

◊

51. A 3.50-kN piano is lifted by three workers at constant speed to an apartment 25.0 m above the street using a pulley system fastened to the roof of the building. Each worker is able to deliver 165 W of power, and the pulley system is 75.0% efficient (so that 25.0% of the mechanical energy is lost due to friction in the pulley). Neglecting the mass of the pulley, find the time required to lift the piano from the street to the apartment.

Solution

The work that must be done on the piano to lift it at constant speed from the street up to the apartment is given by the work-energy theorem as

$$W_{nc} = \Delta KE + \Delta PE = \frac{1}{2}m\left(v_f^2 - v_i^2\right) + mg\left(y_f - y_i\right) = 0 + mg\left(y_f - y_i\right)$$

or

$$W_{nc} = \left(3.50\times 10^3\ \text{N}\right)\left(25.0\ \text{m} - 0\right) = 8.75\times 10^4\ \text{J}$$

The three workmen, working together, deliver a total power input to the pulley system of

$$P_{\text{total}} = 3\left(P_{\substack{\text{single}\\ \text{worker}}}\right) = 3(165\ \text{W}) = 495\ \text{W} = 495\ \text{J/s}$$

However, the pulley system in use is only 75.0% efficient. This means that 25.0% of the power input is used overcoming friction in the system, and only 75.0% of the power input is transferred to the piano as useful work. The rate at which the system does useful work on the piano is

$$P_{\text{useful}} = 0.750P_{\text{total}} = 0.750\left(495\ \text{J/s}\right) = 371\ \text{J/s}$$

The time required to lift the piano is then

$$\Delta t = \frac{W_{nc}}{P_{\text{useful}}} = \frac{8.75\times 10^4\ \text{J}}{371\ \text{J/s}} = 236\ \text{s} = 236\ \text{s}\left(\frac{1\ \text{min}}{60\ \text{s}}\right) = 3.93\ \text{min} \qquad \lozenge$$

57. A 1.50×10^3-kg car starts from rest and accelerates uniformly to 18.0 m/s in 12.0 s. Assume that air resistance remains constant at 400 N during this time. Find (a) the average power developed by the engine and (b) the instantaneous power output of the engine at $t = 12.0$ s, just before the car stops accelerating.

Solution

(a) During the 12.0 s time interval, the car has a constant acceleration of

$$a_x = \frac{\Delta v_x}{\Delta t} = \frac{18.0 \text{ m/s} - 0}{12.0 \text{ s}} = 1.50 \text{ m/s}^2$$

The net force acting on the car is the vector sum of a forward force F_{eng} exerted by the engine (through the drive wheels) and a rearward force f due to air resistance. Newton's second law then becomes $\Sigma F_x = F_{\text{eng}} - f = ma_{\text{max}}$, and gives the forward driving force exerted by the engine as

$$F_{\text{eng}} = f + ma_x = 400 \text{ N} + \left(1.50 \times 10^3 \text{ kg}\right)\left(1.50 \text{ m/s}^2\right) = 2.65 \times 10^3 \text{ N}$$

The average velocity of the car during this time is

$$v_{\text{av}} = \frac{v_f + v_i}{2} = \frac{18.0 \text{ m/s} + 0}{2} = 9.00 \text{ m/s}$$

so the average power input from the engine is

$$P_{\text{eng,av}} = F_{\text{eng}} \cdot v_{\text{av}} = \left(2.65 \times 10^3 \text{ N}\right)\left(9.00 \text{ m/s}\right) = 2.39 \times 10^4 \text{ W}\left(\frac{1 \text{ hp}}{746 \text{ W}}\right) = 32.0 \text{ hp} \qquad \Diamond$$

(b) At $t = 12.0$ s, just before the car ceases to accelerate, the instantaneous velocity of the car is $v = v_f = 18.0$ m/s. Therefore, the instantaneous power input from the engine at this time is

$$P_{\text{eng}} = F_{\text{eng}} \cdot v = \left(2.65 \times 10^3 \text{ N}\right)\left(18.0 \text{ m/s}\right) = 4.77 \times 10^4 \text{ W}\left(\frac{1 \text{ hp}}{746 \text{ W}}\right) = 63.9 \text{ hp} \qquad \Diamond$$

69. (a) A 75-kg man steps out a window and falls (from rest) 1.0 m to a sidewalk. What is his speed just before his feet strike the pavement? (b) If the man falls with his knees and ankles locked, the only cushion for his fall is an approximately 0.50-cm give in the pads of his feet. Calculate the average force exerted on him by the ground during this 0.50 cm of travel. This average force is sufficient to cause cartilage damage in the joints or to break bones.

Solution

(a) While the man is falling, the only force acting on him is the gravitational force (a conservative force). Therefore, the work done by nonconservative forces is zero and the man's mechanical energy (kinetic plus potential) will be conserved. That is,

$$W_{nc} = \left(KE_f - KE_i\right) + \left(PE_f - PE_i\right) = 0 \quad \Rightarrow \quad KE_f + PE_f = KE_i + PE_i$$

Thus,

$$\tfrac{1}{2}mv_f^2 + mgy_f = \tfrac{1}{2}mv_i^2 + mgy_i \quad \text{or} \quad v_f^2 = v_i^2 + 2g\left(y_i - y_f\right)$$

and, when the man starts from rest, his speed just before touching ground (a fall of 1.0 m) will be

$$v_f = \sqrt{v_i^2 + 2g\left(y_i - y_f\right)} = \sqrt{0 + 2\left(9.8\ \text{m/s}^2\right)(1.0\ \text{m})} = 4.4\ \text{m/s} \qquad \Diamond$$

(b) The instant the pads of the man's feet touch ground, the ground starts exerting an upward nonconservative force, F, on him. While this force will probably vary, it will have some average value F_{av} as the pads of the feet are compressed. Consider the period from when the feet first make contact until the man comes to rest, after undergoing an additional displacement of $s = 0.50\ \text{cm} = 5.0 \times 10^{-3}\ \text{m}$ (directed at 180° from the direction of F_{av}). The work done by nonconservative forces is

$$W_{nc} = \left(F_{\text{av}} \cos 180°\right)s = \left(KE_f - KE_i\right) + \left(PE_f - PE_i\right)$$

or

$$-F_{\text{av}}s = \left(0 - \frac{1}{2}mv_i^2\right) + mg\left(y_f - y_i\right)$$

For this time interval, $v_i = 4.4\ \text{m/s}$, the final velocity for the interval in part (a), and $\left(y_f - y_i\right) = -s$. Thus, the magnitude of the average force is

$$F_{\text{av}} = m\left(\frac{v_i^2}{2s} + g\right) = (75\ \text{kg})\left[\frac{(4.4\ \text{m/s})^2}{2\left(5.0 \times 10^{-3}\ \text{m}\right)} + 9.8\ \text{m/s}^2\right] = 1.5 \times 10^5\ \text{N} \qquad \Diamond$$

73. A 2.00×10^2-g particle is released from rest at point A on the inside of a smooth hemispherical bowl of radius $R = 30.0$ cm (Fig. P5.73). Calculate (a) its gravitational potential energy at A relative to B, (b) its kinetic energy at B, (c) its speed at B, and (d) its potential energy at C relative to B, and (e) its kinetic energy at C.

Solution

The smooth bowl exerts no friction force on the particle. Therefore, the only nonconservative force acting on the particle as it slides on the inner surface of the hemispherical bowl is a normal force that is always perpendicular to the motion of the particle (and hence, does no work). This means that the total mechanical energy of the particle (kinetic plus potential energies) remains constant:

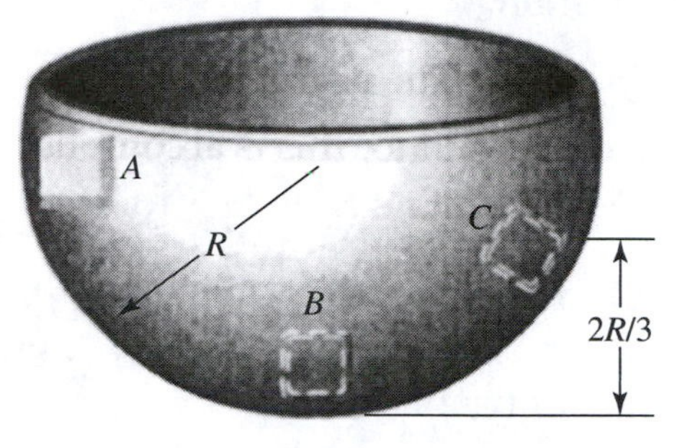

$$W_{nc} = (KE_f - KE_i) + (PE_f - PE_i) = 0 \quad \Rightarrow \quad KE_f + PE_f = KE_i + PE_i$$

If we choose $y = 0$ (and hence, $PE_g = mgy = 0$) at the bottom of the bowl (point B), then $y_A = R = 30.0$ cm and $y_C = 2R/3 = 20.0$ cm $= 0.200$ m.

(a) $(PE_g)_A = mgy_A = (2.00 \times 10^2 \text{ g})\left(\dfrac{1 \text{ kg}}{10^3 \text{ g}}\right)\left(9.80 \dfrac{\text{m}}{\text{s}^2}\right)(30.0 \text{ cm})\left(\dfrac{1 \text{ m}}{10^2 \text{ cm}}\right) = 0.588 \text{ J}$ ◊

(b) In going from A to B, with $v_A = 0$ and $y_B = 0$, we have

$$KE_B + (PE_g)_B = KE_A + (PE_g)_A \quad \text{or} \quad KE_B + 0 = 0 + (PE_g)_A = 0.588 \text{ J} \quad ◊$$

(c) $KE_B = \dfrac{1}{2} m v_B^2 = 0.588$ J

so

$$v_B = \sqrt{\frac{2(0.588 \text{ J})}{m}} = \sqrt{\frac{2(0.588 \text{ J})}{2.00 \times 10^{-1} \text{ kg}}} = 2.42 \text{ m/s} \quad ◊$$

(d) $(PE_g)_C = mgy_C = (2.00 \times 10^{-1} \text{ kg})(9.80 \text{ m/s}^2)(0.200 \text{ m}) = 0.392 \text{ J}$ ◊

(e) As the particle goes from B to C, conservation of energy gives

$$KE_C + (PE_g)_C = KE_B + (PE_g)_B \quad \text{or} \quad KE_C = KE_B + 0 - (PE_g)_C$$

Thus,

$$KE_C = 0.588 \text{ J} - 0.392 \text{ J} = 0.196 \text{ J} \quad ◊$$

79. A ski jumper starts from rest 50.0 m above the ground on a frictionless track and flies off the track at an angle of 45.0° above the horizontal and at a height of 10.0 m above the level ground. Neglect air resistance. (a) What is her speed when she leaves the track? (b) What is the maximum altitude she attains after leaving the track? (c) Where does she land relative to the end of the track?

Solution

We choose a reference level of $y = 0$ (and thus, $PE_g = 0$) at ground level. In the absence of friction or air resistance, mechanical energy is conserved and we may use $KE_f + PE_{g,f} = KE_i + PE_{g,i}$ for any portion of the motion under consideration.

(a) From the moment the skier starts from rest until she leaves the track, conservation of mechanical energy gives $\frac{1}{2}mv_f^2 + mgy_f = 0 + mgy_i$, and her speed as she leaves the end of the track is

$$v_f = \sqrt{2g\left(y_i - y_f\right)} = \sqrt{2\left(9.80 \text{ m/s}^2\right)(50.0 \text{ m} - 10.0 \text{ m})} = 28.0 \text{ m/s} \qquad \lozenge$$

(b) With a speed of $v_i = 28.0$ m/s as she leaves the track, the skier's velocity components at maximum height of the parabolic arc followed while in the air are $v_{fy} = 0$ and $v_{fx} = v_{ix} = v_i \cos 45.0°$. Thus, the speed at maximum height is

$$v_f = \sqrt{v_{fx}^2 + v_{fy}^2} = v_i \cos 45.0°$$

We then obtain $\frac{1}{2}\cancel{m}v_i^2 \cos^2 45.0° + \cancel{m}gy_{\text{max}} = \frac{1}{2}\cancel{m}v_i^2 + \cancel{m}gy_i$ from conservation of mechanical energy. This gives the maximum height attained as

$$y_{\text{max}} = y_i + \frac{v_i^2\left(1 - \cos^2 45.0°\right)}{2g} = 10.0 \text{ m} + \frac{(28.0 \text{ m/s})^2\left(1 - \cos^2 45.0°\right)}{2\left(9.80 \text{ m/s}^2\right)} = 30.0 \text{ m} \qquad \lozenge$$

(c) Applying the kinematics equation $\Delta y = v_{iy}t + \frac{1}{2}a_y t^2$ from when the skier leaves the track to just before she lands gives

$$-10.0 \text{ m} = \left[\left(28.0 \text{ m/s}^2\right)\sin 45.0°\right]t + \tfrac{1}{2}\left(-9.80 \text{ m/s}^2\right)t^2$$

or $t^2 - (4.04 \text{ s})t - 2.04 \text{ s}^2 = 0$. We use the quadratic formula to solve for the time of flight, and find a single positive solution given as $t = 4.49$ s. The horizontal distance traveled beyond the end of the track is then

$$\Delta x = v_{ix}t = \left[(28.0 \text{ m/s})\cos 45.0°\right](4.49 \text{ s}) = 88.9 \text{ m} \qquad \lozenge$$

86. A daredevil wishes to bungee-jump from a hot-air balloon 65.0 m above a carnival midway. He will use a piece of uniform elastic cord tied to a harness around his body to stop his fall at a point 10.0 m above the ground. Model his body as a particle and the cord as having negligible mass and a tension force described by Hooke's force law. In a preliminary test, hanging at rest from a 5.00-m length of the cord, the jumper finds that his body weight stretches it by 1.50 m. He will drop from rest at the point where the top end of a longer section of the cord is attached to the stationary balloon. (a) What length of cord should he use? (b) What maximum acceleration will he experience?

Solution

Each 5.00-m length of the cord will stretch 1.50 m when the tension in the cord equals the weight of the jumper (that is, when $F_s = w = mg$). Thus, the elongation in a cord of original length L when $F_s = w = mg$ will be $x = (L/5.00 \text{ m})(1.50 \text{ m}) = 0.300L$, and the force constant for the cord of length L is

$$k = \frac{F_s}{x} = \frac{mg}{0.300L}$$

(a) In the bungee-jump from the balloon, the daredevil drops $y_i - y_f = 55.0$ m. The stretch of the cord at the start of the jump is $x_i = 0$, and that at the lowest point is $x_f = 55.0 \text{ m} - L$. Since $KE_i = KE_f = 0$ for the fall, conservation of mechanical energy gives $0 + (PE_g)_f + (PE_s)_f = 0 + (PE_g)_i + (PE_s)_i$. This yields $\frac{1}{2}k(x_f^2 - x_i^2) = mg(y_i - y_f)$, or

$$\frac{1}{2}\left(\frac{\cancel{m}\cancel{g}}{0.300L}\right)(55.0 \text{ m} - L)^2 = \cancel{m}\cancel{g}(55.0 \text{ m})$$

which reduces to $(55.0 \text{ m})^2 - (110 \text{ m})L + L^2 = (33.0 \text{ m})L$, or

$$L^2 - (143 \text{ m})L + (55.0 \text{ m})^2 = 0$$

Applying the quadratic formula to this equation yields solutions of $L = 117$ m and $L = 25.8$ m. Considering the fact that the jumper is to leap from an initial height of 65.0 m, the first of these solutions is clearly extraneous and physically unacceptable. Thus, the jumper should choose a cord having a length of 25.8 m. ◊

(b) During the jump, Newton's second law gives

$$\Sigma F_y = kx - mg = ma_y \qquad \text{or} \qquad \left(\frac{\cancel{m}g}{0.300L}\right)x - \cancel{m}g = \cancel{m}a_y$$

Thus, the acceleration of the jumper when the cord is stretched by a distance x is $a_y = (x/0.300L - 1)g$, which is maximum when the jumper is at his lowest point. There, $x = x_{max} = 55.0 \text{ m} - L = 29.2$ m, and the acceleration is

$$(a_y)_{max} = \left[\frac{29.2 \text{ m}}{0.300(25.8 \text{ m})} - 1\right]g = 2.77g = 27.1 \text{ m/s}^2$$ ◊

6

Momentum and Collisions

NOTES FROM SELECTED CHAPTER SECTIONS

6.1 Momentum and Impulse

The **time rate of change of the momentum** of a particle is equal to the resultant force on the particle. The **impulse** of a force is a vector quantity and is equal to the change in momentum of the particle on which the force acts. The impulse or change in momentum of an object is equal to the area under the force vs. time graph from the beginning to the end of the time interval during which the force is in contact with the object. Under the **impulse approximation**, it is assumed that one of the forces acting on a particle is of short time duration but of much greater magnitude than any of the other forces.

6.2 Conservation of Momentum

6.3 Collisions

The principle of **conservation of linear momentum** can be stated for an isolated system of objects. This is a system on which no external forces (e.g., friction or gravity) are acting. **When no external forces act on a system the total linear momentum of the system remains constant.** Remember, momentum is a vector quantity and the momentum of each individual particle may change. The total momentum of the entire system of particles, however, will remain constant. A collision between two or more masses is an important example of conservation of momentum.

For **any type of collision**, the total momentum before the collision equals the total momentum just after the collision.

In an **inelastic collision**, the total momentum is conserved; however, the total kinetic energy is not conserved.

In a **perfectly inelastic** collision, the two colliding objects stick together following the collision. This corresponds to a maximum loss in kinetic energy.

In an **elastic collision**, both momentum and kinetic energy are conserved.

6.4 Glancing Collisions

The law of conservation of momentum is not restricted to one-dimensional collisions. If two masses undergo a **two-dimensional** (glancing) **collision** and there are no external forces acting, the total momentum is conserved in both the x- and y-directions independently.

EQUATIONS AND CONCEPTS

Linear momentum of a particle is defined as the product of its mass m and velocity $\vec{\mathbf{v}}$. The SI units of linear momentum are kg · m/s. Momentum is a vector quantity; the defining equation can be written in component form.

$$\vec{\mathbf{p}} \equiv m\vec{\mathbf{v}} \tag{6.1}$$

$$p_x = mv_x$$

$$p_y = mv_y$$

$$p_z = mv_z$$

The **impulse** imparted to an object is the product of the net force acting on the object and the time interval over which the force acts. *Impulse is a vector quantity and has the same direction as the applied force. The SI unit of impulse is kilogram meter per second.*

$$\vec{\mathbf{I}} \equiv \vec{\mathbf{F}}\Delta t \tag{6.4}$$

The **impulse-momentum theorem** states that the impulse of a force acting on an object equals the change in the momentum of the object. *Impulse is equal to the area under the force-time curve as illustrated by the shaded area in the figure at the right.*

$$\vec{\mathbf{I}} = \vec{\mathbf{F}}_{\text{av}}\,\Delta t = \Delta\vec{\mathbf{p}} = m\vec{\mathbf{v}}_f - m\vec{\mathbf{v}}_i \tag{6.5}$$

The **time-averaged force** is defined as that constant force which would impart the same impulse to a particle as the actual time-varying force acting over the same time interval. *In the figure, the impulse is equal to the area under the dashed line.*

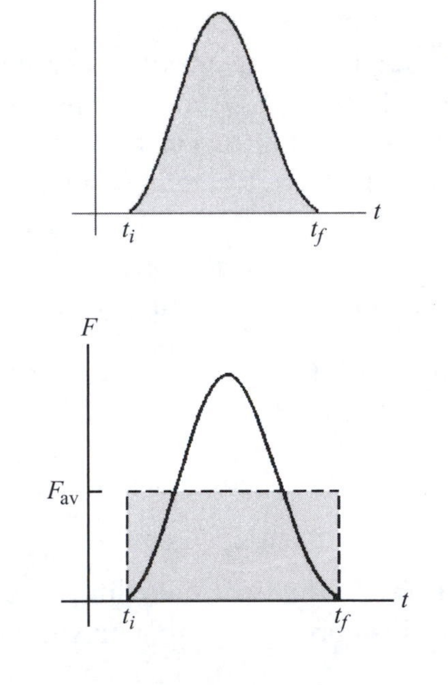

The **principle of conservation of momentum** for a two-body collision is stated in Equation 6.7. *When two objects interact in a collision (and exert internal forces on each other) and no external forces act on the two-object system, the total momentum of the system before the collision equals the total momentum after the collision. Remember, momentum is a vector quantity.*

$$m_1\vec{\mathbf{v}}_{1i} + m_2\vec{\mathbf{v}}_{2i} = m_1\vec{\mathbf{v}}_{1f} + m_2\vec{\mathbf{v}}_{2f} \tag{6.7}$$

$\vec{\mathbf{v}}_{1i}$ $\vec{\mathbf{v}}_{2i}$ m_1 m_2 $\vec{\mathbf{v}}_{1f}$ $\vec{\mathbf{v}}_{2f}$

In a **perfectly inelastic collision**, momentum is conserved but kinetic energy is not. *The colliding objects stick together so that they have a common final velocity.*

$$m_1 v_{1i} + m_2 v_{2i} = (m_1 + m_2) v_f \qquad (6.8)$$

In an **elastic collision**, both momentum and kinetic energy are conserved. *Equations 6.10 and 6.11 apply in the case of "head-on" collisions.*

$$m_1 v_{1i} + m_2 v_{2i} = m_1 v_{1f} + m_2 v_{2f} \qquad (6.10)$$

$$\tfrac{1}{2} m v_{1i}^2 + \tfrac{1}{2} m v_{2i}^2 = \tfrac{1}{2} m_{1f}^2 + \tfrac{1}{2} m_2 v_{2f}^2 \qquad (6.11)$$

The **relative velocity** before a perfectly elastic collision between two bodies equals the negative of the relative velocity of the two bodies following the collision.

$$v_{1i} - v_{2i} = -\,(v_{1f} - v_{2f}) \qquad (6.14)$$

The **final velocities following a one-dimensional elastic collision** between two particles can be calculated when the masses and initial velocities of both particles are known by combining Equations 6.10 and 6.11. *Remember the appropriate algebraic signs (designating direction) must be included for* v_{1i} and v_{2i}.

$$v_{1f} = \left(\frac{m_1 - m_2}{m_1 + m_2}\right) v_{1i} + \left(\frac{2m_2}{m_1 + m_2}\right) v_{2i}$$

$$v_{2f} = \left(\frac{2m_1}{m_1 + m_2}\right) v_{1i} + \left(\frac{m_2 - m_1}{m_1 + m_2}\right) v_{2i}$$

An **important special case** occurs when the second particle (m_2, the "target") is initially at rest.

$$v_{1f} = \left(\frac{m_1 - m_2}{m_1 + m_2}\right) v_{1i}$$

$$v_{2f} = \left(\frac{2m_1}{m_1 + m_2}\right) v_{1i}$$

A **two-dimensional elastic collision** (a glancing collision) in which an object m_1 moves along the x-axis and collides with m_2 initially at rest is illustrated in the figure below. *Momentum is conserved along each direction.* Angles in Equations 6.15 and 6.16 are defined in the diagram.

x-component: (6.15)

$$m_1 v_{1i} + 0 = m_1 v_{1f} \cos\theta + m_2 v_{2f} \cos\phi$$

y-component: (6.16)

$$0 + 0 = m_1 v_{1f} \sin\theta - m_2 v_{2f} \sin\phi$$

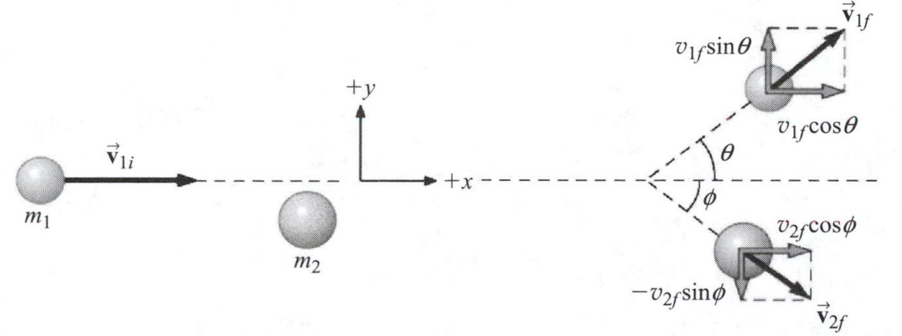

The **equation for rocket propulsion** states that the *change in the speed* of the rocket (as the mass decreases from M_i *to* M_f) is proportional to the exhaust speed of the ejected gases.

$$v_f - v_i = v_e \ln\left(\frac{M_i}{M_f}\right) \qquad (6.19)$$

The **instantaneous thrust on a rocket** is the force exerted on the rocket by the ejected exhaust gases. The thrust increases as both the exhaust velocity and the burn rate increase,

$$\text{Instantaneous Thrust} = \left| v_e \frac{\Delta M}{\Delta t} \right| \qquad (6.20)$$

SUGGESTIONS, SKILLS, AND STRATEGIES

The following procedure is recommended when dealing with problems involving collisions between two objects:

1. Set up a coordinate system and define velocities with respect to that system. That is, objects moving in the direction selected as the positive direction of the x-axis are considered as having a positive velocity and negative if moving in the negative x-direction. It is convenient to have the x-axis coincide with the direction of one of the initial velocities.

2. Show all velocity vectors with labels and include all the given information including scattering angles.

3. Now write expressions for the total momentum before and the total momentum after the collision and equate the two. For two-dimensional collisions, separate expressions must be written for the momentum in the x- and y-directions. See Equations 6.15 and 6.16. *It is important to emphasize that it is the momentum of the system that is conserved, not the momentum of the individual objects.*

4. If the collision is perfectly inelastic (kinetic energy is not conserved and the two objects have a common velocity following the collision), you should then proceed to solve the momentum equations for the unknown quantities.

5. If the collision is perfectly elastic, both momentum and kinetic energy are conserved.

REVIEW CHECKLIST

- The impulse of a force acting on a particle during some time interval equals the change in momentum of the particle, and the impulse equals the area under the force vs. time graph.
- The momentum of any isolated system (or any system for which the net external force is zero) is conserved, regardless of the nature of the forces between the masses that comprise the system.
- There are two types of collisions that can occur between particles, namely elastic and inelastic collisions. Recognize that a perfectly inelastic collision is an inelastic collision in which the colliding particles stick together after the collision, and hence move as a composite particle. Kinetic energy is conserved in elastic collisions, but not in the case of inelastic collisions. Linear momentum is conserved in both elastic and inelastic collisions if the net external force acting on the system of colliding objects is zero.
- The conservation of linear momentum applies not only to head-on collisions (one-dimensional), but also to glancing collisions (in two or three dimensions). For example, in a two-dimensional collision, the total momentum component in the x-direction is conserved and the total momentum component in the y-direction is conserved.
- The equations for conservation of momentum and kinetic energy can be used to calculate the final velocities in a two-body head-on elastic collision. For a perfectly inelastic collision, the equation of conservation of momentum may be used to calculate the final velocity of the composite particle.

SOLUTIONS TO SELECTED END-OF-CHAPTER PROBLEMS

3. A pitcher claims he can throw a 0.145-kg baseball with as much momentum as a 3.00-g bullet moving with a speed of 1.50×10^3 m/s. (a) What must the baseball's speed be if the pitcher's claim is valid? (b) Which has greater kinetic energy, the ball or the bullet?

Solution

(a) If the pitcher's claim is correct, then $p_{\text{ball}} = p_{\text{bullet}}$ or

$$m_{\text{ball}}v_{\text{ball}} = m_{\text{bullet}}v_{\text{bullet}}$$

and the speed of the baseball must be

$$v_{\text{ball}} = \frac{m_{\text{bullet}}v_{\text{bullet}}}{m_{\text{ball}}} = \frac{\left(3.00\times10^{-3}\text{ kg}\right)\left(1.50\times10^{3}\text{ m/s}\right)}{0.145\text{ kg}} = 31.0\text{ m/s} \quad \Diamond$$

(b) The kinetic energy of the bullet is

$$KE_{\text{bullet}} = \frac{1}{2}m_{\text{bullet}}v_{\text{bullet}}^2 = \frac{(3.00 \times 10^{-3}\ \text{kg})(1.50 \times 10^3\ \text{m/s})^2}{2} = 3.38 \times 10^3\ \text{J}$$

and the kinetic energy of the baseball is

$$KE_{\text{ball}} = \frac{1}{2}m_{\text{ball}}v_{\text{ball}}^2 = \frac{(0.145\ \text{kg})(31.0\ \text{m/s})^2}{2} = 69.7\ \text{J}$$

The bullet has the larger kinetic energy by a factor of

$$\frac{KE_{\text{bullet}}}{KE_{\text{ball}}} = \frac{3.38 \times 10^3\ \text{J}}{69.7\ \text{J}} = 48.5$$ ◊

---

9. A 0.280-kg volleyball approaches a player horizontally with a speed of 15.0 m/s. The player strikes the ball with her fist and causes the ball to move in the opposite direction with a speed of 22.0 m/s. (a) What impulse is delivered to the ball by the player? (b) If the player's fist is in contact with the ball for 0.060 0 s, find the magnitude of the average force exerted on the player's fist.

Solution

The sketch at the right shows the volleyball immediately before and immediately after it is hit by the player's fist. Observe that we have arbitrarily chosen the direction of the ball's final velocity to be the positive horizontal direction.

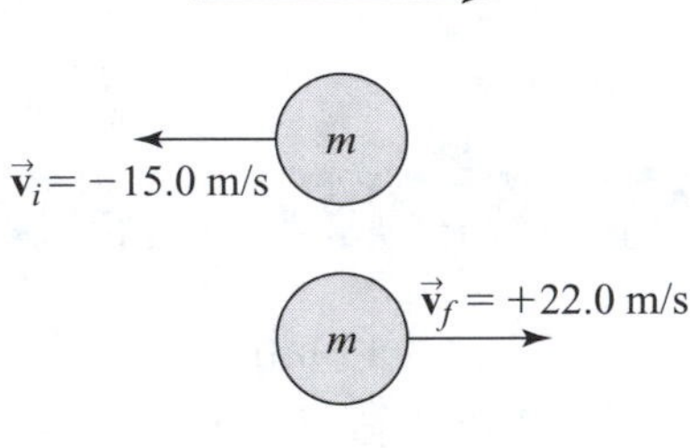

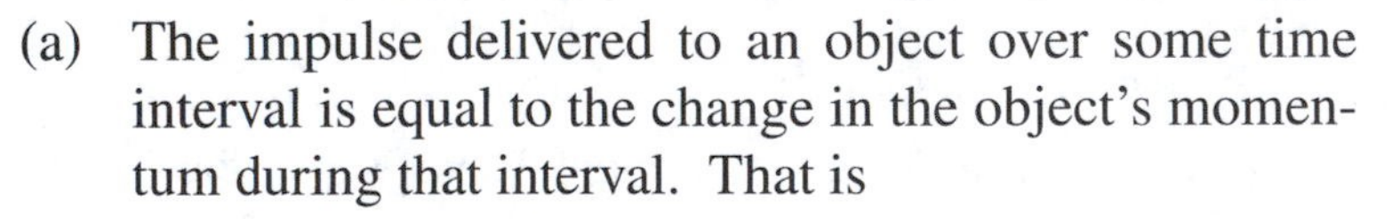

(a) The impulse delivered to an object over some time interval is equal to the change in the object's momentum during that interval. That is

$$\vec{\mathbf{I}} = \Delta\vec{\mathbf{p}} = m\vec{\mathbf{v}}_f - m\vec{\mathbf{v}}_i$$

so

$$\vec{\mathbf{I}} = (0.280\ \text{kg})(+22.0\ \text{m/s}) - (0.280\ \text{kg})(-15.0\ \text{m/s})$$

then

$$\vec{\mathbf{I}} = +6.16\ \text{kg}\cdot\text{m/s} - (-4.20\ \text{kg}\cdot\text{m/s}) = +10.4\ \text{kg}\cdot\text{m/s}$$

or

$\vec{\mathbf{I}} = 10.4\ \text{kg}\cdot\text{m/s}$ in the direction of the ball's final velocity ◊

(b) The impulse delivered to an object in a time interval Δt can also be expressed as $\vec{\mathbf{I}} = \vec{\mathbf{F}}_{av}(\Delta t)$, where $\vec{\mathbf{F}}_{av}$ is the average force exerted on the object during that interval. The average force exerted on the ball during the $\Delta t = 0.600$ s time interval that it is in contact with the player's fist is then

$$\vec{\mathbf{F}}_{av} = \frac{\vec{\mathbf{I}}}{\Delta t} = \frac{+10.4\ \text{kg}\cdot\text{m/s}}{0.060\,0\ \text{s}} = +173\ \text{kg}\cdot\text{m/s}^2 = +173\ \text{N}$$

Thus, the fist exerts an average force on the ball of 173 N in the direction of the ball's final velocity. By Newton's third law, the ball must have exerted an average force of 173 N in the opposite direction (the direction of the ball's initial velocity) on the player's fist. ◊

---

15. The force shown in the force vs. time diagram in Figure P6.15 acts on a 1.5-kg object. Find (a) the impulse of the force, (b) the final velocity of the object if it is initially at rest, and (c) the final velocity of the object if it is initially moving along the x-axis with a velocity of -2.0 m/s.

Solution

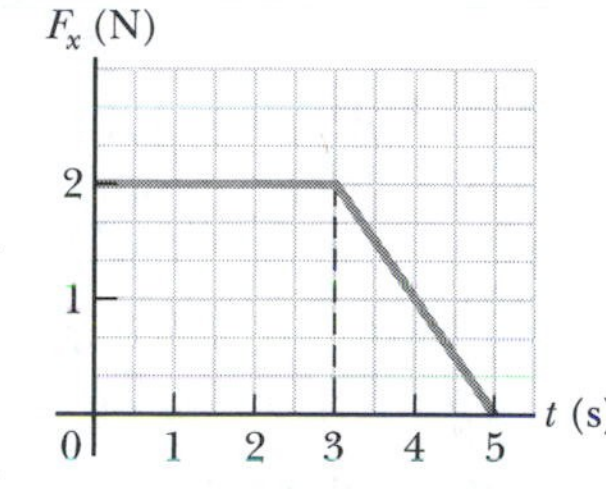

Figure P6.15

(a) The impulse delivered by a force over some time interval is equal to the area under the force versus time graph for that interval. Thus, if the force varies with time as shown in Figure P6.15, the impulse delivered during the interval between $t = 0$ and $t = 5.0$ s is given by the sum of the rectangular area and the triangular area shown. This gives

$$I = A_{\text{rectangle}} + A_{\text{triangle}} = \left[\left(base_{\text{rect}}\right)\left(height_{\text{rect}}\right)\right] + \left[\frac{1}{2}\left(base_{\text{triangle}}\right)\left(height_{\text{triangle}}\right)\right]$$

or

$$I = (3.0\ \text{s})(+2.0\ \text{N}) + \frac{1}{2}(5.0\ \text{s} - 3.0\ \text{s})(+2.0\ \text{N}) = +8.0\ \text{N}\cdot\text{s}$$ ◊

(b) The impulse imparted to an object during a time interval is also equal to the change in momentum the object undergoes during that interval, or $I = \Delta p = mv_f - mv_i$. Thus, if the impulse computed above is imparted to a 1.5-kg object that was initially at rest, the velocity of the object at the end of the 5.0-s interval is

$$v_f = v_i + \frac{\Delta p}{m} = v_i + \frac{I}{m} = 0 + \frac{8.0\ \text{N}\cdot\text{s}}{1.5\ \text{kg}} = +5.3\ \text{m/s}$$ ◊

(c) If the 1.5-kg object was moving along the x-axis with a velocity of $v_i = -2.0$ m/s at $t = 0$, its velocity at time $t = 5.0$ s, after receiving an impulse of $I = +8.0$ N·s, will be

$$v_f = v_i + \frac{I}{m} = -2.0\ \text{m/s} + \frac{8.0\ \text{N}\cdot\text{s}}{1.5\ \text{kg}} = +3.3\ \text{m/s}$$ ◊

---

25. An astronaut in her space suit has a total mass of 87.0 kg, including suit and oxygen tank. Her tether line loses its attachment to her spacecraft while she's on a spacewalk. Initially at rest with respect to her spacecraft, she throws her 12.0-kg oxygen tank away from her spacecraft with a speed of 8.00 m/s to propel herself back toward it (Fig. P6.25). (a) Determine the maximum distance she can be from the craft an return within 2.00 min (the amount of time the air in her helmet remains breathable). (b) Explain in terms of Newton's laws of motion why this strategy works.

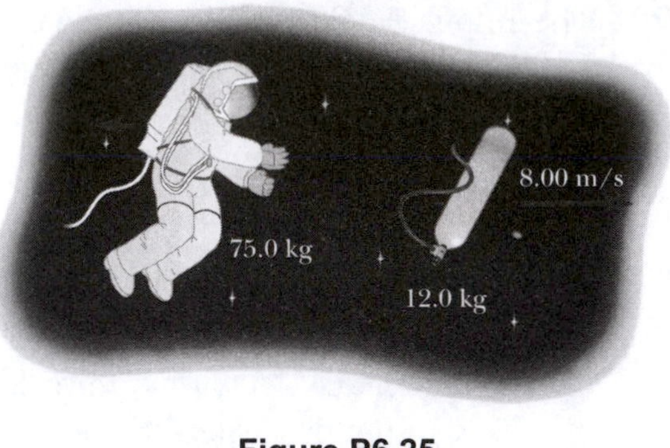

Figure P6.25

Solution

(a) If the oxygen tank has a mass of $m_T = 12.0$ kg, the mass of the astronaut without the tank is $m_A = (87.0\text{ kg} - 12.0\text{ kg}) = 75.0$ kg. Both the astronaut and the tank were initially at rest in a reference frame fixed on the spacecraft. Hence, the initial momentum of a system consisting of the astronaut and the tank was zero in this reference frame. In the absence of external forces acting on the system, total momentum is conserved. This means that the total momentum is still zero after the tank is discarded, or

$$m_A v_{Af} + m_T v_{Tf} = m_A v_{Ai} + m_T v_{Ti} = 0$$

Choosing the direction toward the spacecraft to be the positive direction, the velocity of the astronaut after giving the tank a velocity $v_{Tf} = -8.00\text{ m/s}$ away from the spacecraft is seen to be

$$v_{Af} = -\frac{m_T v_{Tf}}{m_A} = -\frac{(12.0\text{ kg})(-8.00\text{ m/s})}{75.0\text{ kg}} = +1.28\text{ m/s}$$

The distance she will drift toward the spacecraft in the 2.00 minutes her oxygen will last (and therefore the maximum safe distance she can go from the craft) is

$$d = v_{Af}(\Delta t) = (1.28\text{ m/s})(2.00\text{ min})\left(\frac{60.0\text{ s}}{1\text{ min}}\right) = 154\text{ m} \qquad \Diamond$$

(b) From Newton's third law, when the astronaut exerts a force directed away from the spacecraft on the tank, the tank exerts a force of equal magnitude directed toward the craft on her. This force accelerates her toward the ship. ◊

31. Gayle runs at a speed of 4.00 m/s and dives on a sled, initially at rest on the top of a frictionless, snow-covered hill. After she has descended a vertical distance of 5.00 m, her brother, who is initially at rest, hops on her back, and they continue down the hill together. What is their speed at the bottom of the hill if the total vertical drop is 15.0 m? Gayle's mass is 50.0 kg, the sled has a mass of 5.00 kg, and her brother has a mass of 30.0 kg.

Solution

When Gayle dives onto the sled, conservation of momentum gives

$$m_{\text{Gayle+sled}} v_2 = m_{\text{Gayle}} v_1 + m_{\text{sled}} v_0$$

or

$$v_2 = \frac{m_{\text{Gayle}} v_1 + m_{\text{sled}}(0)}{m_{\text{Gayle+sled}}} = \left(\frac{m_{\text{Gayle}}}{m_{\text{Gayle+sled}}}\right) v_1$$

so Gayle and the sled start down the hill with an initial speed of

$$v_2 = \left(\frac{50.0 \text{ kg}}{55.0 \text{ kg}}\right)(4.00 \text{ m/s}) = 3.64 \text{ m/s}$$

After Gayle and the sled have undergone a 5.00 m decrease in altitude, conservation of mechanical energy gives $KE_f = KE_i + (PE_i - PE_f)$, or

$$\frac{1}{2} \cancel{m_{\text{Gayle + sled}}} v_3^2 = \frac{1}{2} \cancel{m_{\text{Gayle + sled}}} v_2^2 + \cancel{m_{\text{Gayle + sled}}} g(y_2 - y_3)$$

and the speed down the incline at this point is

$$v_3 = \sqrt{v_2^2 + 2g(y_2 - y_3)} = \sqrt{(3.64 \text{ m/s})^2 + 2(9.80 \text{ m/s}^2)(5.00 \text{ m})} = \sqrt{111} \text{ m/s}$$

When her brother, initially at rest, drops onto her back, conservation of momentum gives the speed of the combined system down the slope immediately after impact.

$$(m_{\text{Gayle}} + m_{\text{sled}} + m_{\text{brother}}) v_4 = m_{\text{Gayle+ sled}} v_3 + m_{\text{brother}}(0)$$

or

$$v_4 = \left(\frac{m_{\text{Gayle+sled}}}{m_{\text{Gayle}} + m_{\text{sled}} + m_{\text{brother}}}\right) v_3 = \left(\frac{55.0 \text{ kg}}{50.0 \text{ kg} + 5.00 \text{ kg} + 30.0 \text{ kg}}\right)\left(\sqrt{111} \frac{\text{m}}{\text{s}}\right) = 6.82 \text{ m/s}$$

Finally, after the combined system undergoes an additional 10.0 m drop in altitude, conservation of mechanical energy gives $\frac{1}{2} \cancel{m_{\text{total}}} v_5^2 = \frac{1}{2} \cancel{m_{\text{total}}} v_4^2 + \cancel{m_{\text{total}}} g(y_4 - y_5)$ and the speed at the bottom of the hill is

$$v_5 = \sqrt{v_4^2 + 2g(y_4 - y_5)} = \sqrt{(6.82 \text{ m/s})^2 + 2(9.80 \text{ m/s}^2)(10.0 \text{ m})} = 15.6 \text{ m/s}$$ ◊

37. In a Broadway performance, an 80.0-kg actor swings from a 3.75-m-long cable that is horizontal when he starts. At the bottom of his arc, he picks up his 55.0-kg costar in an inelastic collision. What maximum height do they reach after their upward swing?

Solution

The leftmost part of the sketch shows the situation from when the actor starts from rest until just before his impact with his costar. The rightmost part shows the period from just after impact until they come to rest momentarily at the end of the swing.

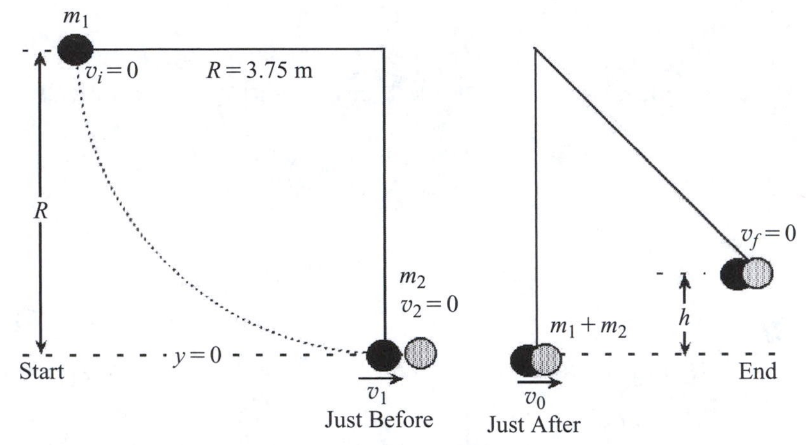

The cable's tension is always perpendicular to the motion and does no work. Thus, mechanical energy is conserved during the actor's downswing before collision. This gives

$$KE_f = KE_i + \left(PE_i - PE_f\right)$$

or

$$\frac{1}{2}\cancel{m_1}v_1^2 = \frac{1}{2}\cancel{m_1}v_0^2 + \cancel{m_1}g\left(y_i - y_f\right)$$

and his speed just before impact is $v_1 = \sqrt{0 + 2gR} = \sqrt{2\left(9.80\ \text{m/s}^2\right)(3.75\ \text{m})} = 8.57\ \text{m/s}$

Conservation of momentum during the collision with his initially stationary costar gives

$$\left(m_1 + m_2\right)v_a = m_1 v_1 + m_2(0)$$

and

$$v_a = \frac{m_1 v_1}{m_1 + m_2} = \frac{(80.0\ \text{kg})(8.57\ \text{m/s})}{80.0\ \text{kg} + 55.0\ \text{kg}} = 5.08\ \text{m/s}$$

is the speed of the pair immediately after the perfectly inelastic collision.

Mechanical energy is again conserved as the actor and costar swing together to height h before coming to rest momentarily. Thus, $KE_f + PE_f = KE_i + PE_i$ or $KE_f + PE_f - PE_i = KE_i$ gives

$$\frac{1}{2}\cancel{\left(m_1 + m_2\right)}(0) + \cancel{\left(m_1 + m_2\right)}g\left(y_f - y_i\right) = \frac{1}{2}\cancel{\left(m_1 + m_2\right)}v_a^2$$

and the maximum height reached after their upward swing is

$$h = \left(y_f - y_i\right) = \frac{v_a^2}{2g} = \frac{(5.08\ \text{m/s})^2}{2\left(9.80\ \text{m/s}^2\right)} = 1.32\ \text{m}$$

◊

46. A billiard ball rolling across a table at 1.50 m/s makes a head-on elastic collision with an identical ball. Find the speed of each ball after the collision (a) when the second ball is initially at rest, (b) when the second ball is moving toward the first at a speed of 1.00 m/s, and (c) when the second ball is moving away from the first at a speed of 1.00 m/s.

Solution

Consider the sketch at the right and apply conservation of momentum from just before to just after the collision to find $mv_{1f} + mv_{2f} = mv_{1i} + mv_{2i}$. Since the two objects have the same mass this reduces to

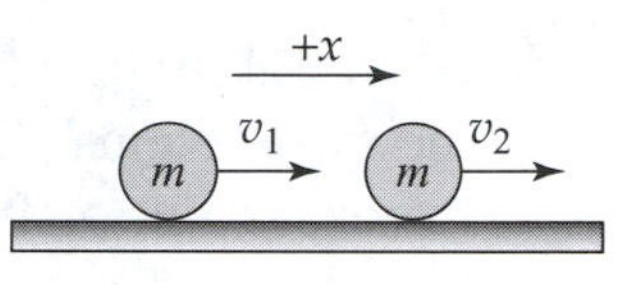

$$v_{1f} + v_{2f} = v_{1i} + v_{2i} \qquad [1]$$

If this head-on collision is also elastic (i.e., a collision in which kinetic energy as well as momentum is conserved), Equation 6.14 from the textbook gives us the relation

$$v_{1i} - v_{2i} = -(v_{1f} - v_{2f})$$

which we may rearrange to obtain

$$-v_{1f} + v_{2f} = v_{1i} - v_{2i} \qquad [2]$$

Adding Equations [1] and [2] gives $v_{2f} = v_{1i}$, while subtracting Equation [2] from Equation [1] yields $v_{1f} = v_{2i}$. Considering these two results, we are drawn to the very interesting conclusion that in an elastic, head-on collision between equal mass objects, the two objects simply swap velocities! That is, the final velocity of one object is always the same as the initial velocity of the other object.

With this observation, we may now simply write down the answers to each of the elastic head-on collisions between equal mass billiard balls in this problem. The results are:

(a) When $v_{1i} = +1.50$ m/s and $v_{2i} = 0$, then $v_{1f} = 0$ and $v_{2f} = +1.50$ m/s. ◊

(b) If $v_{1i} = +1.50$ m/s and $v_{2i} = -1.00$ m/s, the final velocities will be

$$v_{1f} = -1.00 \text{ m/s and } v_{2f} = +1.50 \text{ m/s}$$ ◊

(c) And if $v_{1i} = +1.50$ m/s when $v_{2i} = +1.00$ m/s, the velocities after collision are

$$v_{1f} = +1.00 \text{ m/s and } v_{2f} = +1.50 \text{ m/s}$$ ◊

51. A billiard ball moving at 5.00 m/s strikes a stationary ball of the same mass. After the collision, the first ball moves at 4.33 m/s at an angle of 30° with respect to the original line of motion. (a) Find the velocity (magnitude and direction) of the second ball after collision. (b) Was the collision inelastic or elastic?

Solution

(a) In the absence of any net external force acting on the system consisting of the two balls, total momentum is conserved, or $\vec{\mathbf{p}}_f = \vec{\mathbf{p}}_i$. If this vector equation is to be valid, it is necessary that the x- and y-components be unchanged during the collision, or $\Sigma p_{x,f} = \Sigma p_{x,i}$ and $\Sigma p_{y,f} = \Sigma p_{y,i}$. With $m_1 = m_2 = m$, and $v_{1y,i} = 0$ while $v_{1x,i} = 5.00$ m/s and $v_{1f} = 4.33$ m/s, conserving momentum in the x-direction gives

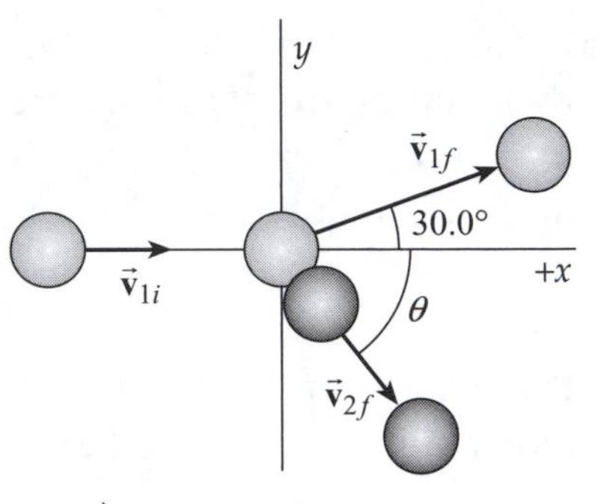

$$\cancel{m}(4.33\text{ m/s})\cos 30.0° + \cancel{m}v_{2f}\cos\phi = \cancel{m}(5.00\text{ m/s}) + 0$$

or

$$v_{2f}\cos\phi = 1.25\text{ m/s} \qquad [1]$$

Likewise, conserving momentum in the y-direction yields

$$\cancel{m}(4.33\text{ m/s})\sin 30.0° + \cancel{m}v_{2f}\sin\phi = 0 + 0$$

or

$$v_{2f}\sin\phi = -2.17\text{ m/s} \qquad [2]$$

Squaring both Equation [1] and Equation [2], and adding the results, gives

$$v_{2f}^2\left(\sin^2\phi + \cos^2\phi\right) = (1.25\text{ m/s})^2 + (-2.17\text{ m/s})^2 = 6.27\text{ m}^2/\text{s}^2$$

Since $\sin^2\phi + \cos^2\phi = 1$, $v_{2f} = \sqrt{6.27}$ m/s $= 2.50$ m/s. We find the direction of $\vec{\mathbf{v}}_{2f}$ by noting that

$$\phi = \tan^{-1}\left(\frac{v_{2y,f}}{v_{2x,f}}\right) = \tan^{-1}\left(\frac{v_{2f}\sin\phi}{v_{2f}\cos\phi}\right) = \tan^{-1}\left(\frac{-2.17\text{ m/s}}{1.25\text{ m/s}}\right) = -60.0°$$

Thus, $\vec{\mathbf{v}}_{2f} = 2.50$ m/s at 60.0° clockwise from the positve x-axis. ◊

(b) The total kinetic energy before impact is

$$KE_i = \frac{1}{2}mv_{1i}^2 + \frac{1}{2}mv_{2i}^2 = \frac{1}{2}m(5.00\text{ m/s})^2 + 0 = m\left(12.5\text{ m}^2/\text{s}^2\right)$$

and the total kinetic energy immediately after impact is

$$KE_f = \frac{1}{2}mv_{1f}^2 + \frac{1}{2}mv_{2f}^2 = \frac{1}{2}m(4.33\text{ m/s})^2 + \frac{1}{2}m(2.50\text{ m/s})^2 = m\left(12.5\text{ m}^2/\text{s}^2\right)$$

Therefore, we see that $KE_f = KE_i$, or kinetic energy as well as momentum has been conserved in the collision. Then, by definition, this is an elastic collision. ◊

57. Two objects of masses $m_1 = 0.56$ kg and $m_2 = 0.88$ kg are placed on a horizontal frictionless surface and a compressed spring of force constant $k = 280$ N/m is placed between them as in Figure P6.57a. Neglect the mass of the spring. The spring is not attached to either object and is compressed a distance of 9.8 cm. If the objects are released from rest, find the final velocity of each object as shown in Figure P6.57b.

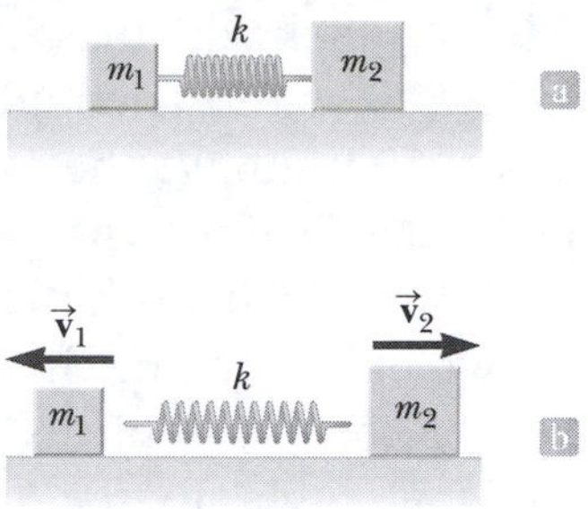

Figure P6.57

Solution

Since the surface is frictionless, both momentum and mechanical energy will be conserved as the spring is released. Also, because the spring is considered massless, it will not possess any momentum or kinetic energy. However, it does possess elastic potential energy when compressed.

Requiring that momentum be conserved gives

$$m_1v_1 + m_2v_2 = 0 + 0$$

or

$$v_1 = -(m_2/m_1)v_2 \qquad \textbf{[1]}$$

If the spring is initially compressed a distance $d = 9.8$ cm, conservation of energy gives $KE_{1f} + KE_{2f} + PE_{s,f} = KE_{1i} + KE_{2,i} + PE_{s,i}$, or $\cancel{\tfrac{1}{2}}m_1v_1^2 + \cancel{\tfrac{1}{2}}m_2v_2^2 + 0 = 0 + 0 + \cancel{\tfrac{1}{2}}kd^2$. Substituting for v_1 from Equation [1] yields $m_1(m_2/m_1)^2 v^2{}_2 + m_2v^2 = kd^2$, which reduces to

$$v_2 = \sqrt{\frac{kd^2}{m_2(1 + m_2/m_1)}}$$

Using the given values for the masses and force constant, the velocity of m_2 after the spring is released is found to be

$$v_2 = \sqrt{\frac{(280 \text{ N/m})(9.8\times10^{-2} \text{ m})^2}{(0.88 \text{ kg})[1 + (0.88 \text{ kg})/(0.56 \text{ kg})]}} = 1.1 \text{ m/s toward the right} \qquad \Diamond$$

Then, Equation [1] gives the final velocity of m_1 as

$$v_1 = -[(0.88 \text{ kg})/(0.56 \text{ kg})](1.1 \text{ m/s}) = -1.7 \text{ m/s} = 1.7 \text{ m/s toward the left} \qquad \Diamond$$

63. A block with mass $m_1 = 0.500$ kg is released from rest on a frictionless track at a distance $h_1 = 2.50$ m above the top of a table. It then collides elastically with an object having mass $m_2 = 1.00$ kg that is initially at rest on the table, as shown in Figure P6.63. (a) Determine the velocities of the two objects just after the collision. (b) How high up the track does the 0.500 kg object travel back after the collision? (c) How far away from the bottom of the table does the 1.00 kg object land, given that the height of the table is $h_2 = 2.00$ m? (d) How far away from the bottom of the table does the 0.500 kg object eventually land?

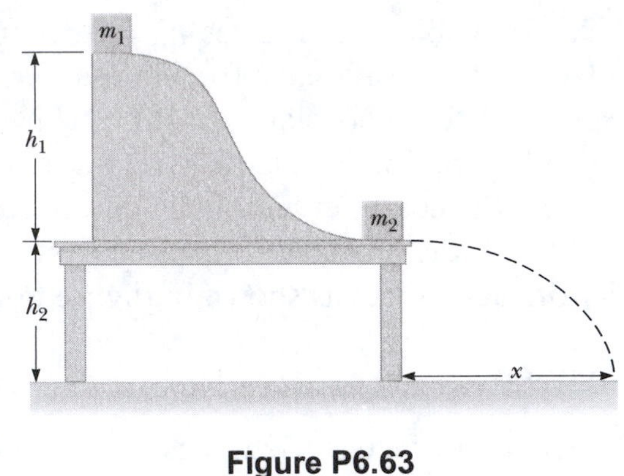

Figure P6.63

Solution

(a) Since the track is frictionless, we use conservation of mechanical energy to find the speed of m_1 just before collision. We find

$$\frac{1}{2}m_1v_1^2 + m_1gy_{\text{bottom}} = \frac{1}{2}m_1v_0^2 + m_1gy_{\text{top}}$$

With $v_0 = 0$, this becomes $v_1 = \sqrt{2g\left(y_{\text{top}} - y_{\text{bottom}}\right)}$, and gives

$$v_1 = \sqrt{2gh_1} = \sqrt{2\left(9.80\ \text{m/s}^2\right)(2.50\ \text{m})} = 7.00\ \text{m/s}$$

From conservation of momentum during the collision and the fact that the velocity of m_2 before impact is $v_2 = 0$, we obtain

$$m_1v_{1f} + m_2v_{2f} = m_1v_1 + m_2v_2$$

or

$$(0.500\ \text{kg})v_{1f} + (1.00\ \text{kg})v_{2f} = (0.500\ \text{kg})(7.00\ \text{m/s}) + 0$$

which simplifies to

$$v_{1f} + 2v_{2f} = 7.00\ \text{m/s} \qquad \textbf{[1]}$$

For elastic head-on collisions such as this, $v_1 - v_2 = -(v_{1f} - v_{2f})$. With $v_2 = 0$, and $v_1 = 7.00$ m/s, this yields

$$v_{1f} - v_{2f} = -7.00\ \text{m/s} \qquad \textbf{[2]}$$

Subtracting Equation [2] from [1] gives $3v_{2f} = 14.0$ m/s and $v_{2f} = 4.67$ m/s ◊

Substituting this result into either Equation [1] or [2] gives $v_{1f} = -2.33$ m/s ◊

The negative sign in the last result means that m_1 rebounds back up the ramp.

(b) After the collision, m_1 rebounds up the ramp, starting with kinetic energy $KE_1' = m_1 v_{1f}^2/2$ at the bottom. It continues up the frictionless ramp until this initial kinetic energy has been converted into gravitational potential energy, or until $m_1 g h_1' = m_1 v_{1f}^2/2$. Thus, the rebound height is

$$h_1' = \frac{v_{1f}^2}{2g} = \frac{(-2.33\ \text{m/s})^2}{2(9.80\ \text{m/s}^2)} = 0.277\ \text{m} \qquad \Diamond$$

(c) From $\Delta y = v_{0y}t + 1/2 a_y t^2$, with $v_{0y} = 0$ and $a_y = -g$, the time required for an object leaving the table horizontally to drop 2.00 m to the floor is

$$t = \sqrt{\frac{2(\Delta y)}{a_y}} = \sqrt{\frac{2(-2.00\ \text{m})}{-9.80\ \text{m/s}^2}} = 0.639\ \text{s}$$

The horizontal distance m_2 travels during its flight to the floor is given by $\Delta x = v_{0x}t$ as

$$x_2 = v_{2f}t = (4.67\ \text{m/s})(0.639\ \text{s}) = 2.98\ \text{m} \qquad \Diamond$$

(d) After rebounding up the ramp to a height $h_1' = 0.277$ m, the 0.500-kg object comes to rest momentarily before sliding back down, converting its energy from gravitational potential energy back into kinetic energy. When it arrives at the bottom, all of its energy is again in the form of kinetic energy, and it leaves the table with a horizontal speed of $v_{1f}' = |v_{1f}| = 2.33$ m/s. As shown in (c) above, the time for it to drop the 2.00 m to the floor is $t = 0.639$ s, and the horizontal distance it travels from the table as it drops is

$$x_1 = v_{1f}'t = (2.33\ \text{m/s})(0.639\ \text{s}) = 1.49\ \text{m} \qquad \Diamond$$

67. A cannon is rigidly attached to a carriage, which can move along horizontal rails, but is connected to a post by a large spring, initially unstretched and with force constant $k = 2.00 \times 10^4$ N/m, as in Figure P6.67. The cannon fires a 200-kg projectile at a velocity of 125 m/s directed 45.0° above the horizontal. (a) If the mass of the cannon and its carriage is 5 000 kg, find the recoil speed of the cannon. (b) Determine the maximum extension of the spring. (c) Find the maximum force the spring exerts on the carriage. (d) Consider the system consisting of the cannon, carriage, and shell. Is the momentum of this system conserved during the firing? Why or why not?

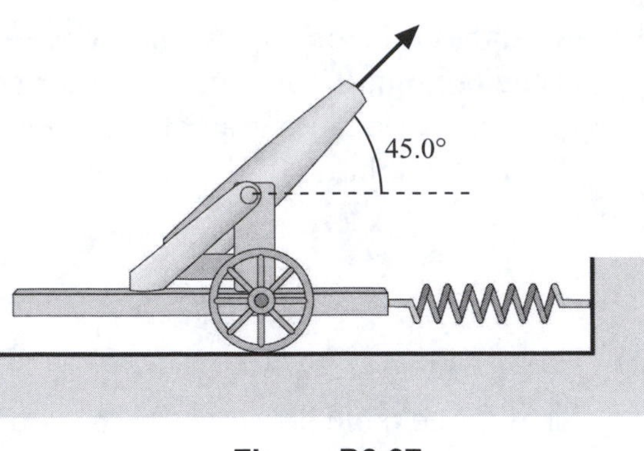

Figure P6.67

Solution

(a) The spring does not have time to stretch during the very brief interval from just before the cannon firing to the instant just after firing. Thus, without any horizontal, *external* forces acting on the system (cannon + spring + shell) during this time interval, the horizontal component of momentum must be conserved. From this, and taking toward the right as positive, we obtain

$$(\Sigma p_x)_f = (\Sigma p_x)_i$$

or

$$m_{\text{shell}}\left(v_{\text{shell}} \cos 45.0°\right) + m_{\text{cannon}} v_{\text{recoil}} = 0 + 0$$

which gives the recoil velocity of the cannon as

$$v_{\text{recoil}} = -\left(\frac{m_{\text{shell}}}{m_{\text{cannon}}}\right) v_{\text{shell}} \cos 45.0° = -\left(\frac{200 \text{ kg}}{5\,000 \text{ kg}}\right)(125 \text{ m/s})\cos 45.0° = -3.54 \text{ m/s}$$

and the recoil speed is

$$|v_{\text{recoil}}| = 3.54 \text{ m/s} \qquad \Diamond$$

(b) As the cannon recoils and the spring stretches, no external forces do work on the cannon plus spring system. Therefore, we make use of conservation of mechanical energy to obtain

$$\left(KE + PE_g + PE_s\right)_f = \left(KE + PE_g + PE_s\right)_i$$

or

$$0 + 0 + \frac{1}{2}kx_{\text{max}}^2 = \frac{1}{2}m_{\text{cannon}}\, v_{\text{recoil}}^2 + 0 + 0$$

and find

$$x_{\max} = \sqrt{\frac{m_{\text{cannon}} v_{\text{recoil}}^2}{k}} = \sqrt{\frac{(5\,000 \text{ kg})(-3.54 \text{ m/s})^2}{2.00 \times 10^4 \text{ N/m}}} = 1.77 \text{ m}$$ ◊

(c) The spring exerts maximum force (directed toward the right) when it has the maximum extension of $x_{\max}$. The magnitude of this maximum force is

$$|F_{\max}| = k x_{\max} = (2.00 \times 10^4 \text{ N/m})(1.77 \text{ m}) = 3.54 \times 10^4 \text{ N}$$ ◊

(d) The vertical component of the system's (cannon + spring + shell) momentum *is not conserved* during the firing of the cannon. The reason for this is that the rail exerts a vertical external force (the normal force) on the cannon and prevents it from recoiling in the vertical direction as the shell starts up the cannon's barrel. However, the spring does not have time to stretch during the cannon firing. Thus, the spring does not yet exert a force on the post, and the post does not yet exert a horizontal reaction force on the system. Without any horizontal external forces present, *the horizontal component of the system's momentum is conserved* during the brief instant from just before to just after the firing. This was the principle we made use of in part (a).

---

75. *Measuring the speed of a bullet.* A bullet of mass m is fired horizontally into a wooden block of mass M lying on a table. The bullet remains in the block after the collision. The coefficient of friction between the block and table is μ, and the block slides a distance d before stopping. Find the initial speed v_0 of the bullet in terms of M, m, μ, g, and d.

Solution

Consider the sketches at the right showing the situation just before the bullet strikes the block, immediately after impact, and at the end with the block at rest after sliding distance d across the table.

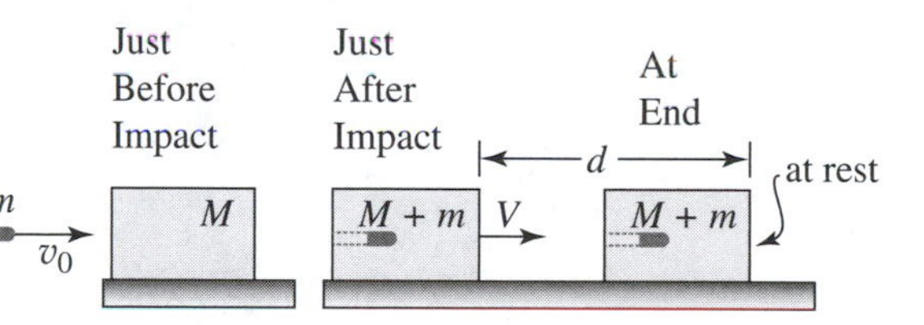

At first glance, one might be tempted to use conservation of energy from before impact all the way to the end with the block at rest again. However, during the collision, an unknown amount of the bullet's initial kinetic energy is "lost" to internal energy as the bullet splits wood fibers, is deformed and heated, and possibly even melts. Thus, we use a different approach by using conservation of momentum through the difficult period from just before to just after impact (but before the block has had a chance to move). This gives

$$mv_0 + 0 = (M + m)V$$

or

$$v_0 = [(M/m) + 1]V \qquad \textbf{[1]}$$

To obtain an expression for the velocity of the (block + embedded bullet) immediately after impact, we can now use the work-energy theorem in the form $W_{\text{net}} = KE_f - KE_i$. This is possible because the "loss" of an unknown amount of kinetic energy is now past and it is possible to account for whatever mechanical energy remains. The normal force exerted on the block + bullet by the table is $n = (M+m)g$, so the friction force has magnitude $f = \mu n = \mu(M+m)g$. The work done by this force as the block slides is $W_{nc} = (f\cos 180°)d = -fd$. Therefore, the work-energy theorem becomes

$$-\mu\cancel{(M+m)}gd = 0 - \tfrac{1}{2}\cancel{(M+m)}V^2$$

and yields $V = \sqrt{2\mu gd}$. Substituting this result into Equation [1] gives the desired expression for the speed of the bullet. This is

$$v_0 = \left[(M/m)+1\right]\sqrt{2\mu gd} = \left(\frac{M+m}{m}\right)\sqrt{2\mu gd}$$ ◊

7

Rotational Motion and the Law of Gravity

NOTES FROM SELECTED CHAPTER SECTIONS

7.1 Angular Speed and Angular Acceleration

Pure rotational motion refers to the motion of a rigid body about a fixed axis. In the case of **rotation about a fixed axis**, every particle on the rigid body has the same angular velocity and the same angular acceleration.

One **radian** (rad) is the angle subtended by an arc length equal to the radius of the arc. That is, the angle measured in radians is given by the arc length divided by the corresponding radius. One complete rotation of 360 degrees equals 2π radians.

7.2 Rotational Motion Under Constant Angular Acceleration

The equations for rotational motion under constant angular acceleration are of the same form as those for linear motion under constant linear acceleration with the substitutions $x \to \theta$, $v \to \omega$, and $a \to \alpha$.

7.3 Relations Between Angular and Linear Quantities

When a **rigid body rotates about a fixed axis**, every point in the object moves along a circular path which has its center at the axis of rotation. The instantaneous velocity of each point is directed along a tangent to the circle. Every point on the object experiences the *same angular speed*; however, points that are different distances from the axis of rotation have different tangential speeds. The value of each of the linear quantities — displacement (s), tangential speed (v_t), and tangential acceleration (a_t) — is equal to the radial distance from the axis multiplied by the corresponding angular quantity, θ, ω, and α.

7.4 Centripetal Acceleration

In circular motion, the centripetal acceleration is directed inward toward the center of the circle and has a magnitude given either by v^2/r or $r\omega^2$. The force that causes centripetal acceleration acts toward the center of the circular path about which the object rotates. If the force vanishes, the object does not continue to move in its circular path; instead, it moves along a straight-line path tangent to the circle.

7.5 Newtonion Gravitation

There are several important features of the law of universal gravitation.

The gravitational force

- is an action-at-a-distance force that always exists between two particles regardless of the medium that separates them,
- varies as the inverse square of the distance between the particles and therefore decreases rapidly with increasing separation, and
- is proportional to the product of their masses.

The gravitational force exerted by a uniform spherical mass on a particle outside the sphere is the same as if the entire mass of the sphere were concentrated at the center.

7.6 Kepler's Laws

Kepler's laws applied to the solar system are:

1. All planets move in elliptical orbits with the Sun at one of the focal points.
2. A line drawn from the Sun to any planet sweeps out equal areas in equal time intervals.
3. The square of the orbital period of any planet is proportional to the cube of the average distance from the planet to the Sun.

EQUATIONS AND CONCEPTS

Arc length, *s*, is the distance traveled by a particle as it moves along a circular path of radius *r*. The radial line from the center of the path to the particle sweeps out an angle, θ (measured counterclockwise relative to a reference direction). The **radian**, a unit of angular measure, is the ratio of two lengths (arc length to radius) and hence is a dimensionless quantity.

$$\theta = \frac{s}{r} \qquad (7.1)$$

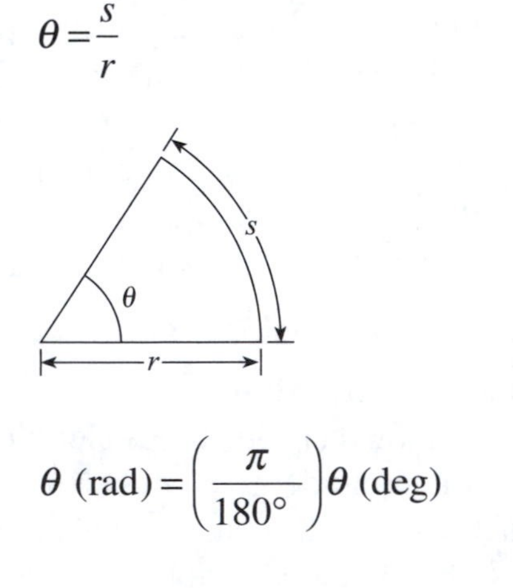

$$\theta\ (\text{rad}) = \left(\frac{\pi}{180°}\right)\theta\ (\text{deg})$$

$$1\ \text{rad} = 57.3°$$

The **average angular speed** of a rotating object is the ratio of the angular displacement (measured in radians) to the time interval during which the angular displacement occurs.

$$\omega_{\text{av}} \equiv \frac{\theta_f - \theta_i}{t_f - t_i} = \frac{\Delta\theta}{\Delta t} \qquad (7.3)$$

The **average angular acceleration** of a rotating object is the ratio of the change in angular velocity to the time interval during which the change in velocity occurs.

$$\alpha_{\text{av}} \equiv \frac{\omega_f - \omega_i}{t_f - t_i} = \frac{\Delta\omega}{\Delta t} \qquad (7.5)$$

The **equations of rotational kinematics** describe the motion of a particle or extended body rotating about a fixed axis with *constant acceleration.* Note that the rotational equations, involving the angular variables $\Delta\theta$, ω, and α, have a one-to-one correspondence with the equations of linear motion, involving the variables Δx, v, and a.

$$\omega = \omega_i + \alpha t \qquad (7.7)$$

$$v = v_i + at$$

$$\Delta\theta = \omega_i t + \tfrac{1}{2}\alpha t^2 \qquad (7.8)$$

$$\Delta x = v_i t + \tfrac{1}{2}at^2$$

$$\omega^2 = \omega_i^2 + 2\alpha\,\Delta\theta \qquad (7.9)$$

$$v^2 = v_i^2 + 2a\Delta x$$

The **magnitudes of the tangential velocity** and **tangential acceleration** of a given point on a rotating object are related to the corresponding angular quantities via the radius of the path along which the point moves. The tangential velocity and tangential acceleration are directed along the tangent to the circular path (and therefore perpendicular to the radius from the center of rotation). The figure shows a point on a disk rotating counterclockwise with *negative* angular acceleration. *Every point on a rotating object has the same value of ω and the same value of α.*

$$v_t = r\omega \qquad (7.10)$$

$$a_t = r\alpha \qquad (7.11)$$

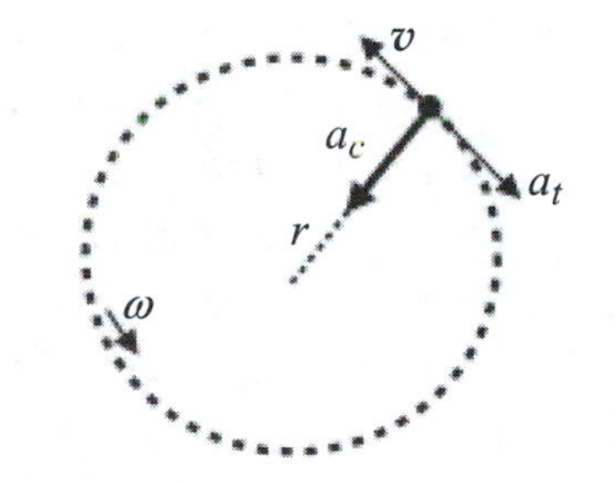

The **centripetal acceleration** of an object in circular motion is directed toward the center of the circle (see figure on the following page) with a magnitude that depends on the values of the tangential (or angular) velocity and the radius of the path.

$$a_c = \frac{v^2}{r} \qquad (7.13)$$

or

$$a_c = r\omega^2 \qquad (7.17)$$

The **total acceleration** of an object in rotational motion has two components as shown in the accompanying diagram.

$$a = \sqrt{a_t^2 + a_c^2} \tag{7.18}$$

$$\theta = \tan^{-1}\left(\frac{a_t}{a_c}\right)$$

The **tangential component** (a_t) given by Equation 7.11 is directed along the direction of travel.

The **centripetal component** (a_c) given by Equation 7.13 is directed perpendicular to the direction of travel.

The **total acceleration** (a) (magnitude and direction) can be found by using Equation 7.18 and the equation for the angle θ.

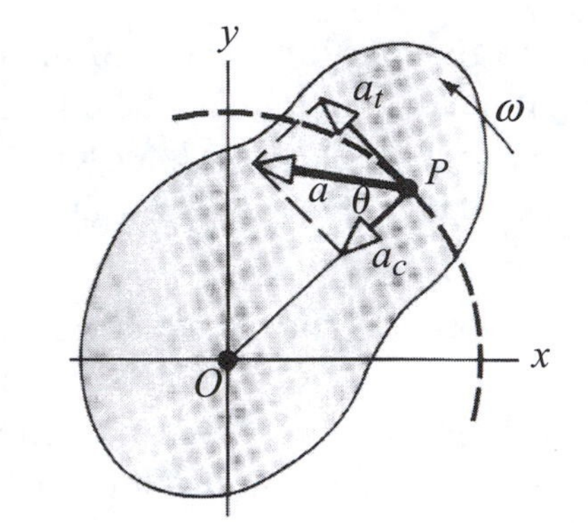

A net **radial (or centripetal)** force is required to maintain motion along a circular path and is always directed toward the center of the path. *Equation 7.19 is a statement of Newton's second law along the radial direction.*

$$F_r = ma_c = m\frac{v^2}{r} \tag{7.19}$$

The **law of universal gravitation** states that every particle in the universe attracts every other particle with a force that is directly proportional to the product of their masses and inversely proportional to the square of the distance between them. The constant G is called the universal gravitational constant. *The magnitudes of the forces of gravitational attraction on the two masses shown in the figure are equal regardless of the relative values of m_1 and m_2.*

$$F = G\frac{m_1 m_2}{r^2} \tag{7.20}$$

$$G = 6.673 \times 10^{-11}\ \text{kg}^{-1} \cdot \text{m}^3 \cdot \text{s}^{-2}$$

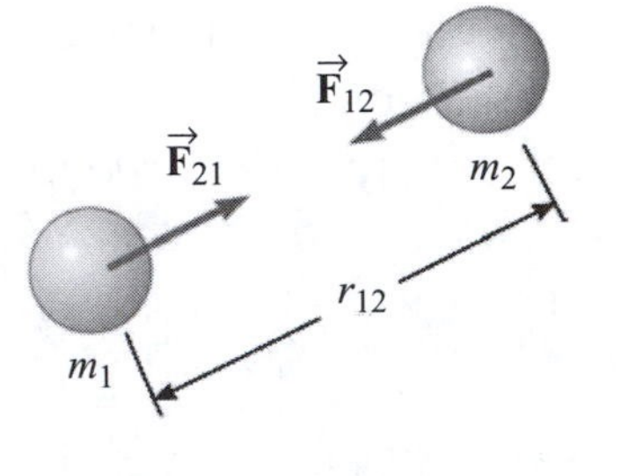

The **gravitational potential energy** associated with a mass m at a distance r above the center of the Earth is given by Equation 7.21. Note that PE will be *negative* for a non-infinite value of r.

$$PE = -G\frac{M_E m}{r} \tag{7.21}$$

The **escape speed** is the minimum speed that an object projected upward from the Earth's surface must have in order to escape from the Earth's gravitational field. *The escape speed does not depend on the mass of the object.*

$$v_{esc} = \sqrt{\frac{2GM_E}{R_E}} \tag{7.22}$$

Kepler's third law states that the square of the orbital period of a planet is proportional to the cube of the mean distance from the planet to the Sun. The value of K_S is independent of the mass of the planet orbiting the Sun.

$$T^2 = \left(\frac{4\pi^2}{GM_S}\right) r^3 = K_S r^3 \tag{7.23}$$

$$K_S = \frac{4\pi^2}{GM_S} = 2.97 \times 10^{-19} \text{ s}^2/\text{m}^3$$

For an Earth satellite, (e.g., the Moon orbiting the Earth) M_S in Equation 7.23 must be replaced by M_E, the mass of the Earth, and the constant K would have a different value.

SUGGESTIONS, SKILLS, AND STRATEGIES

When solving problems involving rotational motions and centripetal accelerations, the following points should be kept in mind:

1. Draw a free-body diagram of the object(s) under consideration, showing all forces that act on it (them).

2. Choose a coordinate system with one axis along the radial direction (i.e., tangent to the path followed by the object) and the other axis perpendicular to the radial direction.

3. Find the net force toward the center of the circular path. This is the centripetal force, the force which causes the centripetal acceleration.

4. From this point onward, the steps are virtually identical to those encountered when solving Newton's second law problems with $F_x = ma_x$ and $F_y = ma_y$. In this case, Newton's second law is applied along the radial (directed toward the center) and tangential directions. Also, you should note that the magnitude of the centripetal acceleration can always be written for circular motion as $a_c = v^2/r$.

REVIEW CHECKLIST

- Quantitatively, angular displacement, angular velocity, and angular acceleration for a rigid body system in rotational motion are related to linear displacement, tangential velocity, and tangential acceleration, respectively. Each linear quantity is calculated by multiplying the corresponding angular quantity by the radius for that particular object or point.

- If a body rotates about a fixed axis, every particle on the body has the same angular velocity and angular acceleration. For this reason, rotational motion can be simply described using these quantities. The formulas which describe angular motion are analogous to the corresponding set of equations pertaining to linear motion.

- A particle moving along a circular path with constant speed experiences an acceleration although the magnitude of the velocity remains constant. This is true because the direction of the velocity (always tangent to the circular path) changes with time. The direction of the acceleration is toward the center of the circular path.

- When both the magnitude and direction of $\vec{\mathbf{v}}$ are changing with time, there are two components of acceleration for a particle moving on a curved path. In this case, the particle has a tangential component of acceleration and a radial component of acceleration.

- Newton's law of universal gravitation is an example of an inverse-square law, and it describes an attractive force between two particles separated by a distance r.

SOLUTIONS TO SELECTED END-OF-CHAPTER PROBLEMS

3. The tires on a new compact car have a diameter of 2.0 ft and are warrantied for 60 000 miles. (a) Determine the angle (in radians) through which one of these tires will rotate during the warranty period. (b) How many revolutions of the tire are equivalent to your answer in part (a)?

Solution

(a) If the tire does not slip on the roadway, a point on the tread of the tire must travel 1 mile in its circular path around the axle for each mile the car travels down the road. That is, the point on the tread always undergoes a displacement around its circular path that is equal to the linear displacement of the car if there is no slipping of the tire on the pavement. Thus, we may use the relation $s = r\theta$, between linear displacement s and angular displacement θ, to find the angle through which one of these tires rotates during the warranty period. This gives

$$\theta = \frac{s}{r} = \frac{60\ 000 \text{ miles}}{1.0 \text{ ft}}\left(\frac{5\ 280 \text{ ft}}{1 \text{ mile}}\right) = 3.2 \times 10^{8} \text{ radians} \quad \Diamond$$

(b) If there is no slipping of the tire on the roadway, the car advances a distance equal to the circumference of the tire for every revolution of the tire. Therefore, the number of revolutions the tire makes during the 60 000-mile warranty period is

$$n = \frac{60\ 000 \text{ miles}}{\textit{circumference}} = \frac{60\ 000 \text{ miles}}{2\pi r} = \frac{60\ 000 \text{ miles}}{2\pi(1.0 \text{ ft})}\left(\frac{5\ 280 \text{ ft}}{1 \text{ mile}}\right) = 5.0 \times 10^{7} \text{ rev} \quad \Diamond$$

7. A machine part rotates at an angular speed of 0.06 rad/s; its speed is then increased to 2.2 rad/s at an angular acceleration of 0.70 rad/s^2. (a) Find the angle through which the part rotates before reaching this final speed. (b) In general, if both the initial and final angular speed are doubled at the same angular acceleration, by what factor is the angular displacement changed? Why? [*Hint:* Look at the form of Equation 7.9.]

Solution

(a) Just as was the case in applying kinematics equations to solve problems in uniformly accelerated linear motion, there are generally several approaches one can use in applying kinematics equations to solve problems in uniformly accelerated rotational motion. One approach we can use to solve this problem is to use the definition of angular acceleration ($\alpha = \Delta\omega/\Delta\tau$) to solve for the time required to produce the desired change in angular speed. This yields

$$t - 0 = \frac{\Delta\omega}{\alpha} = \frac{\omega_f - \omega_i}{\alpha} = \frac{2.2 \text{ rad/s} - 0.06 \text{ rad/s}}{0.70 \text{ rad/s}^2} = 3.1 \text{ s}$$

Then, we can use either $\Delta\theta = \bar{\omega} \cdot t = \left[(\omega_f + \omega_i)/2\right] \cdot t$ or $\Delta\theta = \omega_i t + \frac{1}{2}\alpha t^2$ to find the angular displacement during this time interval. Using the first of these equations gives

$$\Delta\theta = \left(\frac{\omega_f + \omega_i}{2}\right) \cdot t = \left(\frac{2.2 \text{ rad/s} + 0.06 \text{ rad/s}}{2}\right)(3.1 \text{ s}) = 3.5 \text{ rad} \qquad \Diamond$$

We can obtain the same result by employing the equation $\omega_f^2 = \omega_i^2 + 2\alpha(\Delta\theta)$ without having to first solve for the elapsed time. This yields

$$\Delta\theta = \frac{\omega_f^2 - \omega_i^2}{2\alpha} = \frac{(2.2 \text{ rad/s})^2 - (0.06 \text{ rad/s})^2}{2(0.70 \text{ rad/s}^2)} = 3.5 \text{ rad} \qquad \Diamond$$

(b) Considering the last approach used to solve for the angular displacement above, we note that the relation $\omega_f^2 = \omega_i^2 + 2\alpha(\Delta\theta)$ involves the *squares* of both the initial and final angular velocities. Thus, if both of these velocities are doubled (i.e., increased by a factor of 2), the resulting angular displacement will be increased by a factor of $2^2 = 4$, or will be quadrupled. ◊

14. An electric motor rotating a workshop grinding wheel at a rate of 1.00×10^2 rev/min is switched off. Assume the wheel has a constant negative angular acceleration of magnitude 2.00 rad/s^2. (a) How long does it take for the grinding wheel to stop? (b) Through how many radians has the wheel turned during the interval found in part (a)?

Solution

(a) The initial angular velocity of the grinding wheel is given to be

$$\omega_i = 1.00 \times 10^2 \text{ rev/min} = \left(1.00 \times 10^2 \frac{\text{rev}}{\text{min}}\right)\left(\frac{2\pi \text{ rad}}{1 \text{ rev}}\right)\left(\frac{1 \text{ min}}{60.0 \text{ s}}\right) = 10.5 \text{ rad/s}$$

From the definition of angular acceleration, $\alpha = \Delta\omega/\Delta t$, the time required for an object to undergo a change $\Delta\omega$ in its angular velocity is $\Delta t = (\omega_f - \omega_i)/\alpha$. In this case, the final angular velocity is to be $\omega_f = 0$ (it comes to rest). Thus, if $\alpha = -2.00 \text{ rad/s}^2$, the required time is

$$\Delta t = \frac{\Delta\omega}{\alpha} = \frac{0 - (10.5 \text{ rad/s})}{-2.00 \text{ rad/s}^2} = 5.25 \text{ s}$$ ◊

(b) The angular displacement an object with constant angular acceleration will undergo in time Δt is given by $\Delta\theta = \bar{\omega} \cdot \Delta t$, where $\bar{\omega} = (\omega_f + \omega_i)/2$. Thus, the angular displacement of the grinding wheel as it slows and comes to rest is

$$\Delta\theta = \left(\frac{0 + 10.5 \text{ rad/s}}{2}\right)(5.25 \text{ s}) = 27.6 \text{ rad}$$ ◊

This result could also have been found from the kinematics relation $\Delta\theta = \omega_i(\Delta t) + \frac{1}{2}\alpha(\Delta t)^2$ as follows:

$$\Delta\theta = (10.5 \text{ rad/s})(5.25 \text{ s}) + \frac{1}{2}\left(-2.00 \text{ rad/s}^2\right)(5.25 \text{ s})^2 = 27.6 \text{ rad}$$ ◊

19. One end of a cord is fixed and a small 0.500-kg object is attached to the other end, where it swings in a section of a vertical circle of radius 2.00 m, as shown in Figure P7.19. When $\theta = 20.0°$, the speed of the object is 8.00 m/s. At this instant, find (a) the tension in the string, (b) the tangential and radial components of acceleration, and (c) the total acceleration. (d) Is your answer changed if the object is swinging down toward its lowest point instead of swinging up? (e) Explain your answer to part (d).

Solution

(a) The sketch at the right shows the forces acting on the object when the string makes an angle θ with the vertical. The net force acting along the radius line and directed toward the center of the circular path must produce the needed centripetal acceleration. Thus, from Newton's second law, we have

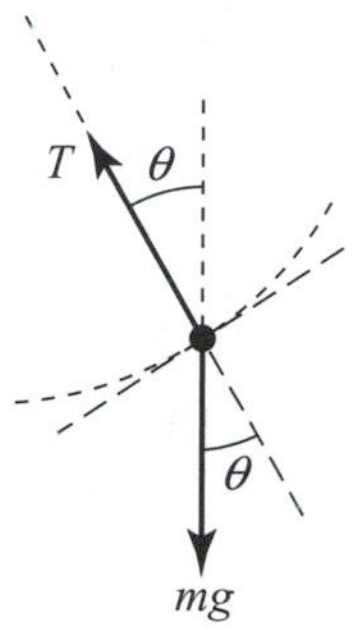

$$\Sigma F_c = T - mg\cos\theta = ma_c$$

When the string is at angle $\theta = 20.0°$ and the 0.500-kg object is moving at speed $v = 8.00$ m/s along a circular path of radius $r = 2.00$ m, the tension in the string will be

$$T = ma_c + mg\cos\theta = m\left(\frac{v^2}{r} + g\cos\theta\right)$$

$$= (0.500\text{ kg})\left(\frac{(8.00\text{ m/s})^2}{2.00\text{ m}} + (9.80\text{ m/s}^2)\cos 20.0°\right) = 20.6\text{ N} \quad \Diamond$$

(b) As seen in the above sketch, the net force tangent to the circular path is $\Sigma F_t = mg\sin\theta$. Thus, from Newton's second law, the tangential acceleration at the instant when $\theta = 20.0°$ is down and toward the left with magnitude

$$a_t = \frac{\Sigma F_t}{m} = \frac{mg\sin\theta}{m} = g\sin\theta = (9.80\text{ m/s}^2)\sin 20.0° = 3.35\text{ m/s}^2 \quad \Diamond$$

and, if $v = 8.00$ m/s at this instant, the radial (or centripetal) acceleration is

$$a_c = \frac{v^2}{r} = \frac{(8.00\text{ m/s})^2}{2.00\text{ m}} = 32.0\text{ m/s}^2 \quad \Diamond$$

(c) The total acceleration then has a magnitude of

$$a_{\text{total}} = \sqrt{a_t^2 + a_c^2} = \sqrt{(3.35\text{ m/s}^2)^2 + (32.0\text{ m/s}^2)^2} = 32.2\text{ m/s}^2$$

This acceleration is oriented to the left of the string (radius line) at an angle of

$$\phi = \tan^{-1}\left(\frac{a_t}{a_c}\right) = \tan^{-1}\left(\frac{3.35 \text{ m/s}^2}{32.0 \text{ m/s}^2}\right) = 5.98°$$

or at angle $\theta + \phi = 20.0° + 5.98° = 26.0°$ from the vertical.

$$\vec{\mathbf{a}}_{\text{total}} = 32.2 \text{ m/s}^2 \text{ at } 26.0° \text{ to the left of vertical}$$ ◊

(d) In no part of the calculation of the total acceleration given above have we specified whether the object was swinging upward or back down toward its lowest point. Hence, the result should be independent of which way the object is swinging. ◊

(e) The tension in the string is the only nonconservative force acting on the object, and this force is always perpendicular to the motion. Hence, it does no work on the object and the total mechanical energy of the object (kinetic energy plus gravitational potential energy) is constant. Since the object always has the same gravitational potential energy at this location (with the string at 20.0° from the vertical), the kinetic energy, and hence the speed, must also always have the same values at this location. This means that the tension force, and therefore, the total force acting on the object is the same each time the object comes to this location. Therefore, the total acceleration $\vec{\mathbf{a}}_{\text{total}} = \vec{\mathbf{F}}_{\text{total}}/m$ must always be equal to the result of part (c) each time the string is at 20.0° from the vertical regardless of whether the object is swinging upward or swinging downward. ◊

---

27. An air puck of mass $m_1 = 0.25$ kg is tied to a string and allowed to revolve in a circle of radius $R = 1.0$ m on a frictionless horizontal table. The other end of the string passes through a hole in the center of the table, and a mass of $m_2 = 1.0$ kg is tied to it (Fig. P7.27). The suspended mass remains in equilibrium while the puck on the tabletop revolves. (a) What is the tension in the string? (b) What is the horizontal force acting on the puck? (c) What is the speed of the puck?

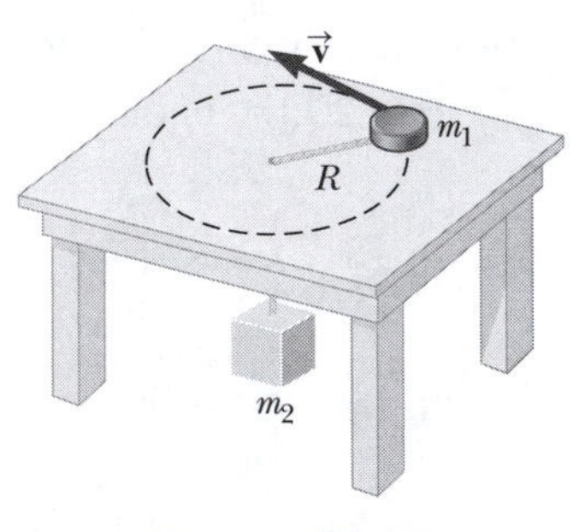

Figure P7.27

Solution

(a) Since the suspended mass on the end of the string is in equilibrium, the resultant force acting on it must be zero. The forces acting on this object are the upward directed tension in the string and the downward gravitational force. Therefore,

$$\Sigma F_y = T - m_2 g = 0$$

or $$T = m_2 g = (1.0 \text{ kg})(9.80 \text{ m/s}^2) = 9.8 \text{ N}$$ ◊

(b) The horizontal force acting on the puck is the tension in the portion of the string that is above the table. Assuming that the string is free to slide, without friction, through the hole in the tabletop, the tension is uniform throughout the length of the string. In this case the magnitude of the horizontal force acting on the puck is $T = 9.8$ N as found in part (a) above. ◊

(c) The puck, moving in a circular path of radius r, has a centripetal acceleration $a_c = v^2/r$. The net force acting on the puck and directed toward the center of the circular path (the tension in the string in this case) must produce this acceleration. Newton's second law then gives $F_c = T = m_1 v^2/r$, or

$$v = \sqrt{\frac{Tr}{m_1}} = \sqrt{\frac{(9.8\ \text{N})(1.0\ \text{m})}{0.25\ \text{kg}}} = 6.3\ \text{m/s} \quad \lozenge$$

---

36. After the Sun exhausts its nuclear fuel, its ultimate fate may be to collapse to a *white dwarf* state. In this state, it would have approximately the same mass as it has now, but its radius would be equal to the radius of Earth. Calculate (a) the average density of the white dwarf, (b) the surface free-fall acceleration, and (c) the gravitational potential energy associated with a 1.00-kg object at the surface of the white dwarf.

Solution

(a) The average density of an object is the ratio of the mass of the object to its volume (i.e., the mass per unit volume). For a spherical object, this is

$$\rho_{\text{av}} = \frac{M}{V} = \frac{M}{4\pi r^3/3} = \frac{3M}{4\pi r^3}$$

When the Sun collapses to a white dwarf state, its mass will be $M \approx M_S = 1.991\times10^{30}$ kg and its radius will be $r \approx R_E = 6.38\times10^6$ m. (Note that these values and similar data on some members of our solar system are given in Table 7.3 of the textbook.) The average density of the white dwarf will be

$$\rho_{\text{av}} = \frac{3\left(1.991\times10^{30}\ \text{kg}\right)}{4\pi\left(6.38\times10^6\ \text{m}\right)^3} = 1.83\times10^9\ \text{kg/m}^3 \quad \lozenge$$

(b) Newton's law of universal gravitation gives the gravitational force acting on an object of mass m on the surface of the white dwarf as $F_g = mg = G M_{\text{Sun}} m/R_E^2$. Therefore, the free-fall acceleration on the surface is

$$g = \frac{F_g}{m} = \frac{GM}{R_E^2} = \frac{\left(6.67\times10^{-11}\ \text{N}\cdot\text{m}^2/\text{kg}^2\right)\left(1.991\times10^{30}\ \text{kg}\right)}{\left(6.38\times10^6\ \text{m/s}\right)^2} = 3.26\times10^6\ \text{m/s}^2 \quad \lozenge$$

(c) Outside of a spherical distribution of matter having mass M, the gravitational potential energy of an object of mass m located distance r from the center is $PE_g = -GMm/r$. The gravitational potential energy of a 1.00-kg mass on the surface of the white dwarf is then

$$PE_g = -\frac{GM_{\text{Sun}}m}{R_E} = -\frac{(6.67\times 10^{-11}\ \text{N}\cdot\text{m}^2/\text{kg}^2)(1.991\times 10^{30}\ \text{kg})(1.00\ \text{kg})}{6.38\times 10^6\ \text{m}}$$

or $PE_g = -2.08\times 10^{13}\ \text{J}$ ◊

---

42. An artificial satellite circling the Earth completes each orbit in 110 minutes. (a) Find the altitude of the satellite. (b) What is the value of g at the location of this satellite?

Solution

(a) When a satellite orbits a central body of mass M, Kepler's third law gives the relation between its orbital period T and the semimajor axis a of its orbit. This relation is

$$T^2 = \left(\frac{4\pi^2}{GM}\right)a^3$$

where $G = 6.67\times 10^{-11}\ \text{N}\cdot\text{m}^2/\text{kg}^2$ is the universal gravitation constant. In this case, the central body is the Earth with mass $M = M_E = 5.98\times 10^{24}$ kg. The period of the orbital motion is $T = 110\ \text{min} = 6.60\times 10^3$ s; and, for a circular orbit, the semimajor axis is simply the radius ($a = r$). Therefore, the radius of the orbit of this satellite must be

$$r = \left[\frac{GM_E T^2}{4\pi^2}\right]^{1/3} = \left[\frac{(6.67\times 10^{-11}\ \text{N}\cdot\text{m}^2/\text{kg}^2)(5.98\times 10^{24}\ \text{kg})(6.60\times 10^3\ \text{s})^2}{4\pi^2}\right]^{1/3}$$

which gives $r = 7.61\times 10^6$ m. The altitude of the satellite above the surface of Earth is then

$$h = r - R_E = 7.61\times 10^6\ \text{m} - 6.38\times 10^6\ \text{m} = 1.23\times 10^6\ \text{m}$$ ◊

(b) From Newton's law of universal gravitation, the gravitational force exerted on an object of mass m by the Earth (mass M_E) when that object is located at distance r from the center of Earth is

$$F_g = \frac{GM_E m}{r^2}, \text{ where } G = 6.67\times 10^{-11}\ \text{N}\cdot\text{m}^2/\text{kg}^2$$

The acceleration of gravity at the location of a satellite in an orbit of radius $r = 7.61\times10^6$ m around the Earth is then

$$g = \frac{F_g}{m} = \frac{GM_E}{r^2} = \frac{\left(6.67\times10^{-11}\ \text{N}\cdot\text{m}^2/\text{kg}^2\right)\left(5.98\times10^{24}\ \text{kg}\right)}{\left(7.61\times10^6\ \text{m}\right)^2} = 6.89\ \text{m/s}^2 \quad \Diamond$$

---

49. One method of pitching a softball is called the "windmill" delivery method, in which the pitcher's arm rotates through approximately 360° in a vertical plane before the 198-gram ball is released at the lowest point of the circular motion. An experienced pitcher can throw a ball with a speed of 98.0 mi/h. Assume that the angular acceleration is uniform throughout the pitching motion and take the distance between the softball and the shoulder joint to be 74.2 cm. (a) Determine the angular speed of the arm in rev/s at the instant of release. (b) Find the value of the angular acceleration in rev/s² and the radial and tangential acceleration of the ball just before it is released. (c) Determine the force exerted on the ball by the pitcher's hand (both radial and tangential components) just before it is released.

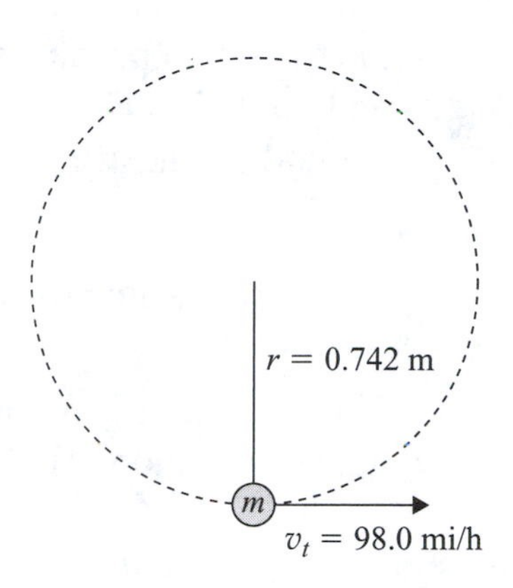

Solution

(a) The ball, at the end of the arm, has a tangential speed of $v_t = 98.0$ mi/h just before it is released at the lowest point on the circular path. At this instant, the ball, hand, and arm have an angular speed of

$$\omega = \frac{v_t}{r} = \frac{98.0\ \text{mi/h}}{0.742\ \text{m}}\left(\frac{1\ \text{m/s}}{2.237\ \text{mi/h}}\right)\left(\frac{1\ \text{rev}}{2\pi\ \text{rad}}\right) = 9.40\ \text{rev/s} \quad \Diamond$$

(b) Assuming the ball has had a uniform angular acceleration during the entire 360° or 1 revolution of the pitching motion, the angular acceleration is given by the rotational kinematics relation $\omega^2 = \omega_i^2 + 2\alpha(\Delta\theta)$ as

$$\alpha = \frac{\omega^2 - \omega_i^2}{2(\Delta\theta)} = \frac{(9.40\ \text{rev/s})^2 - 0}{2(1\ \text{rev})} = 44.2\ \text{rev/s}^2 \quad \Diamond$$

The tangential acceleration of the ball is $a_t = r\alpha$, where α is expressed in units of rad/s². Therefore,

$$a_t = r\alpha = (0.742\ \text{m})\left[\left(44.2\ \frac{\cancel{\text{rev}}}{\text{s}^2}\right)\left(\frac{2\pi\ \text{rad}}{1\ \cancel{\text{rev}}}\right)\right] = 206\ \text{m/s}^2 \quad \Diamond$$

The radial, or centripetal, acceleration of the ball at the moment of release is

$$a_c = \frac{v_t^2}{r} = \frac{\left[(98.0 \text{ mi/h})\left(\frac{1 \text{ m/s}}{2.237 \text{ mi/h}}\right)\right]^2}{0.742 \text{ m}} = 2.59 \times 10^3 \text{ m/s}^2 \quad \Diamond$$

(c) The sketch at the right is a force diagram of the ball at the moment of release. The net force directed toward the center of the circular path is the force that produces the centripetal acceleration. From Newton's second law, $F_c = F_r - mg = ma_c$. The radial component of force exerted on the ball by the arm is then

$$F_r = ma_c + mg = m(a_c + g) = (0.198 \text{ kg})(2.59 \times 10^3 \text{ m/s}^2 + 9.80 \text{ m/s}^2) = 515 \text{ N} \quad \Diamond$$

The tangential component of force the arm exerts on the ball gives the ball its tangential acceleration. This component is given by

$$F_t = ma_t = (0.198 \text{ m/s}^2)(206 \text{ m/s}^2) = 40.8 \text{ N} \quad \Diamond$$

---

53. The Solar Maximum Mission Satellite was placed in a circular orbit about 150 mi above Earth. Determine (a) the orbital speed of the satellite and (b) the time required for one complete revolution.

Solution

(a) If the Solar Maximum Mission Satellite orbited at an altitude of $h = 150$ mi above the surface of the Earth, the radius of its orbit was

$$r = R_E + h = 6.38 \times 10^6 \text{ m} + (150 \text{ mi})\left(\frac{1609 \text{ m}}{1 \text{ mi}}\right) = 6.62 \times 10^6 \text{ m}$$

The gravitational force exerted on a satellite by the central body it orbits supplies the required centripetal acceleration to hold the satellite in orbit. That is, $\Sigma F_c = F_g = ma_c = m(v^2/r)$. Thus, the orbital speed of this satellite must have been

$$v = \sqrt{\frac{F_g \cdot r}{m}} = \sqrt{\frac{(GM_E m/r^2) \cdot r}{m}} = \sqrt{\frac{GM_E}{r}}$$

or $$v = \sqrt{\frac{(6.67 \times 10^{-11} \text{ N} \cdot \text{m}^2/\text{kg}^2)(5.98 \times 10^{24} \text{ kg})}{6.62 \times 10^6 \text{ m}}} = 7.76 \times 10^3 \text{ m/s} \quad \Diamond$$

which is approximately 17 400 miles per hour.

(b) During each revolution around the Earth, the satellite traveled a distance equal to the circumference of its orbit ($C = 2\pi r$). Its period, or the time required to complete this trip, was then

$$T = \frac{2\pi r}{v} = \frac{2\pi(6.62 \times 10^6 \text{ m})}{7.76 \times 10^3 \text{ m/s}} = 5.36 \times 10^3 \text{ s}\left(\frac{1 \text{ min}}{60.0 \text{ s}}\right) = 89.3 \text{ min} \quad \lozenge$$

---

57. Because of Earth's rotation about its axis, a point on the equator experiences a centripetal acceleration of $0.034\,0 \text{ m/s}^2$, while a point at the poles has no centripetal acceleration. (a) Show that, at the equator, the gravitational force on an object (the object's true weight) must exceed the object's apparent weight. (b) What are the apparent weights of a 75.0-kg person at the equator and at the poles? (Assume Earth is a uniform sphere, and take $g = 9.800 \text{ m/s}^2$.)

Solution

(a) The sketch at the right is a view from above the north pole of Earth, showing an object on the equator being weighed with a spring scale. Note that this object is moving in a circular path of radius $r = R_E$ with the angular speed of the rotating Earth. The scale reading (the apparent weight of the object) is equal to the upward normal force exerted on the object by the scale. The true weight of the object is equal to the downward gravitational force exerted on the object by the Earth.

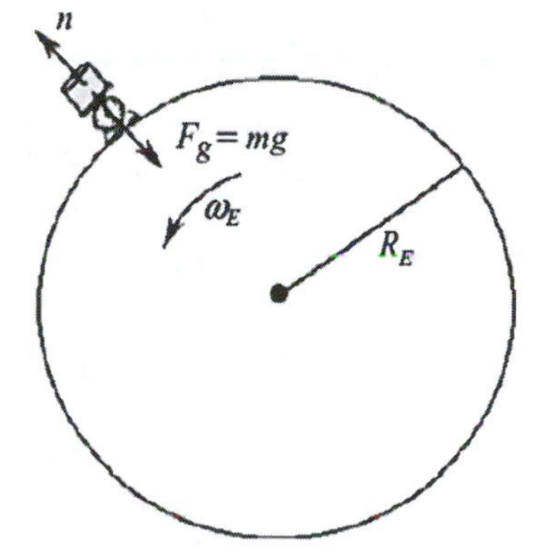

The net force directed toward the center of the circular path must supply the centripetal acceleration. Thus, $\Sigma F_{\substack{\text{toward}\\\text{center}}} = F_g - n = ma_c$ and the true weight is

$$F_g = n + ma_c = \textit{apparent weight} + ma_c > \textit{apparent weight} \quad \lozenge$$

(b) From above, the apparent weight of an object is $n = F_g - ma_c = m(g - a_c)$. The centripetal acceleration of an object on the rotating Earth is $a_c = r\omega_E^2$, where r is the radius of the circular path the object follows around Earth's axis.

At the equator, $r = R_E$, and it is given that $a_c = R_E\omega_E^2 = 0.034\,0 \text{ m/s}^2$. The apparent weight of a 75.0-kg person at the equator is

$$n = m(g - a_c) = (75.0 \text{ kg})(9.800 \text{ m/s}^2 - 0.034\,0 \text{ m/s}^2) = 732 \text{ N} \quad \lozenge$$

At the north pole, the radius of the circular path followed by the person is $r = 0$ and the centripetal acceleration is $a_c = r\omega_E^2 = 0$. The apparent weight of the person here is

$$n = m(g - a_c) = (75.0 \text{ kg})(9.80 \text{ m/s}^2 - 0) = 735 \text{ N} = \text{ the true weight} \quad \lozenge$$

---

65. Suppose a 1 800-kg car passes over a bump in a roadway that follows the arc of a circle of radius 20.4 m, as in Figure P7.65. (a) What force does the road exert on the car as the car passes the highest point of the bump if the car travels at 8.94 m/s? (b) What is the maximum speed the car can have without losing contact with the road as it passes this highest point?

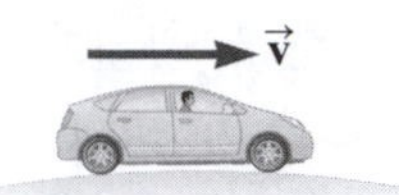

Figure P7.65

Solution

(a) The force diagram at the right shows the forces acting on the car as it passes over the highest point on the bump of radius r. The upward normal force n is the force exerted on the car by the road. The downward gravitational force is the weight of the car. The net force directed downward toward the center of the circular path of the car must supply the needed centripetal acceleration if the car is to stay in contact with the roadway.

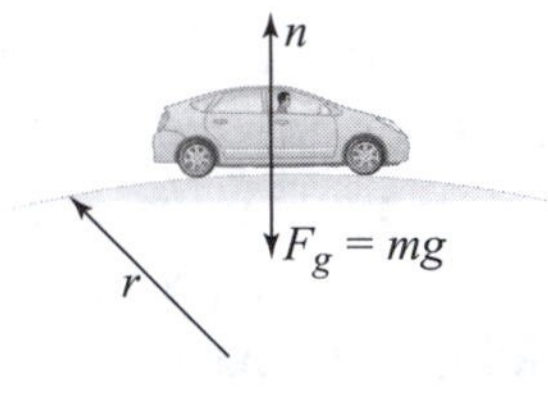

Applying Newton's second law, taking downward as positive, gives

$$\Sigma F_c = F_g - n = mg - n = ma_c$$

and since the centripetal acceleration may be written as $a_c = v^2/r$, the normal force exerted on the car by the road when it passes over the hump with a speed of $v = 8.94\text{ m/s}$ becomes

$$n = m\left(g - \frac{v^2}{r}\right) = (1\,800\text{ kg})\left(9.80\text{ m/s}^2 - \frac{(8.94\text{ m/s})^2}{20.4\text{ m}}\right)$$

or

$$n = 1.06 \times 10^4\text{ N} = 10.6\text{ kN} \qquad \lozenge$$

(b) Observe from above that the normal force decreases as the speed of the car increases. At the critical speed $v = \sqrt{rg}$, the normal force goes to zero. The car will leave the roadway as it crests the bump at any higher speed. The critical speed in this case is

$$v = \sqrt{rg} = \sqrt{(20.4\text{ m})(9.80\text{ m/s}^2)} = 14.1\text{ m/s} \qquad \lozenge$$

70. A 0.275-kg object is swung in a *vertical* circular path on a string 0.850 m long as in Figure P7.70. (a) What are the forces acting on the ball at any point along this path? (b) Draw free-body diagrams for the ball when it is at the bottom of the circle and when it is at the top. (c) If its speed is 5.20 m/s at the top of the circle, what is the tension in the string there? (d) If the string breaks when its tension exceeds 22.5 N, what is the maximum speed the object can have at the bottom before the string breaks?

Solution

(a) At any point on the circular path, the ball has two forces acting on it. One is the gravitational force, with the constant magnitude $F_g = mg$ and is always directed vertically downward. The other is the tension in the string, which is always directed inward toward the center of the circular path, but varies in magnitude from point to point on the path. ◊

(b) The sketch at the right includes two free-body diagrams, one of the ball when at the bottom of the circular path, and one when the ball is at the highest point on the path. Note how the tension force is directed toward the center in both cases, while the gravitational force is always directed downward. The vector sum of these forces must provide the needed centripetal acceleration at each point on the path.

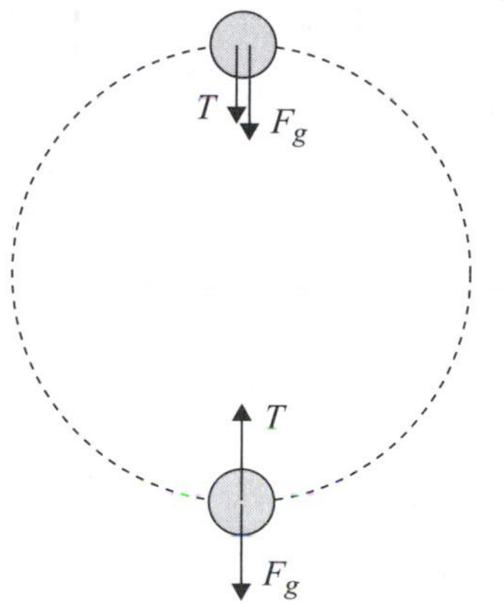

(c) At the top of the circle, the net force directed toward the center is $F_c = T + mg = mv^2/r$. The tension in the string at this point on the path must be

$$T = m\left(\frac{v^2}{r} - g\right) = (0.275 \text{ kg})\left(\frac{(5.20 \text{ m/s})^2}{0.850 \text{ m}} - 9.80 \text{ m/s}^2\right) = 6.05 \text{ N} \quad ◊$$

(d) At the lowest point on the circular path, the net force toward the center is $F_c = T - mg = mv^2/r$. Thus, if the tension in the string at this point is to be $T = T_{max}$, the speed of the ball must be $v = \sqrt{(r/m)(T_{max} - mg)}$. With $T_{max} = 22.5$ N, this speed is

$$v = \sqrt{\left(\frac{0.850 \text{ m}}{0.275 \text{ kg}}\right)\left[22.5 \text{ N} - (0.275 \text{ kg})(9.80 \text{ m/s}^2)\right]} = 7.82 \text{ m/s} \quad ◊$$

8

Rotational Equilibrium and Rotational Dynamics

NOTES FROM SELECTED CHAPTER SECTIONS

8.1 Torque

Torque is the physical quantity that is a measure of the tendency of a force to cause rotation of a body about a specified axis. **Torque is a vector quantity and must be defined with respect to a specific axis of rotation**. Torque, which has the SI **units** of $N \cdot m$, must not be confused with work and energy.

8.2 Torque and the Two Conditions for Equilibrium

A body in static equilibrium must satisfy two conditions:

1. The resultant external force must be zero.

2. The resultant external torque about any axis must be zero.

8.3 The Center of Gravity

In order to calculate the torque due to the weight (gravitational force) on a rigid body, the entire weight of the object can be considered to be concentrated at a single point called the center of gravity. *The center of gravity of a homogeneous, symmetric body must lie along an axis of symmetry*

8.5 Relationship Between Torque and Angular Acceleration

The angular acceleration of an object is proportional to the net torque acting on it. The moment of inertia of the object is the proportionality constant between the net torque and the angular acceleration. **The force and mass in linear motion correspond to torque and moment of inertia in rotational motion**. Moment of inertia of an object depends on the location of the axis of rotation and upon the manner in which the mass is distributed relative to that axis (e.g., a ring has a greater moment of inertia than a disk of the same mass and radius).

8.6 Rotational Kinetic Energy

In linear motion, the energy concept is useful in describing the motion of a system. The energy concept can be equally useful in simplifying the analysis of rotational motion. We now have expressions for four types of mechanical energy: gravitational potential energy, PE_g; elastic potential energy, PE_s; translational kinetic energy, KE_t; and rotational kinetic energy, KE_r. We must include all these forms of energy in the equation for conservation of mechanical energy.

EQUATIONS AND CONCEPTS

The magnitude and direction of the torque about a specified axis depend on:

- the magnitude of the applied force
- the point of application of the force
- the direction of the applied force

$$\tau = rF\sin\theta \quad (8.2)$$

The **magnitude of the torque** is given by Equation 8.2. Either θ or θ' as shown in the figure on the right may be used in Equation 8.2.

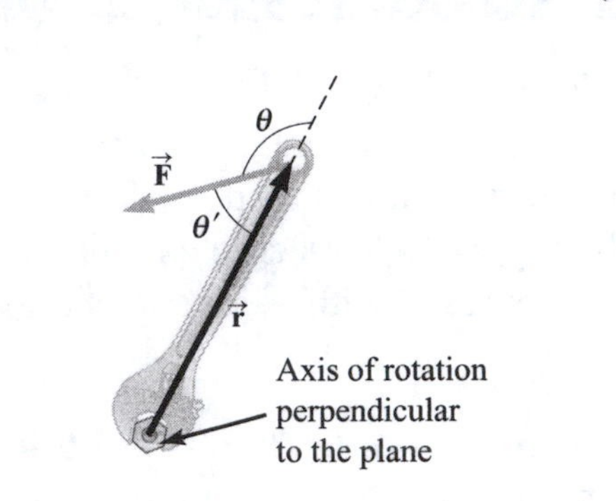

The **direction of the torque** will be out of the plane (positive) or into the plane (negative) when the force and the position vectors are perpendicular to the axis of rotation.

$\vec{\mathbf{r}}$ is the vector from the axis of rotation to the point of application of the force.

The three-part **right-hand rule** stated at the right can be used to determine the direction of the torque vector.

1. Point the fingers of your right hand in the direction of $\vec{\mathbf{r}}$.
2. Curl your fingers toward the direction of $\vec{\mathbf{F}}$.
3. Your thumb will point in the direction of the torque.

The **algebraic sign of a torque: another example**. Torque is considered positive if the applied force has a tendency to rotate the body counterclockwise about the chosen axis, and negative if the tendency for rotation is clockwise. In the figure at the right with the axis perpendicular to the page at point O:
The torque due to $\vec{\mathbf{F}}_1$ is positive.
The torque due to $\vec{\mathbf{F}}_2$ is negative.

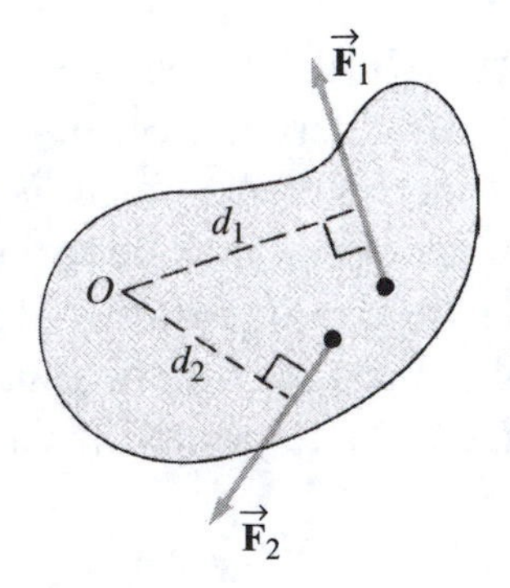

The **first condition for equilibrium** requires that the net external force acting on a body equal zero.

$\Sigma\vec{\mathbf{F}} = 0$

Translational equilibrium

The **second condition for equilibrium** requires that the net external torque acting on a body equals zero.

$\Sigma\vec{\tau} = 0$

Rotational equilibrium

The **center of gravity** is the point where the total weight of an object can be considered to be concentrated when computing the torque due to the force of gravity (an object's weight). The coordinates of the center of gravity are x_{cg}, y_{cg}, and z_{cg}. *The center of gravity of a symmetric homogeneous body must lie on its axis of symmetry but not necessarily on the object.*

$$x_{cg} = \frac{\Sigma m_i x_i}{\Sigma m_i} \qquad (8.3a)$$

$$y_{cg} = \frac{\Sigma m_i y_i}{\Sigma m_i} \qquad (8.3b)$$

$$z_{cg} = \frac{\Sigma m_i z_i}{m_i} \qquad (8.3c)$$

The **angular acceleration** of a point mass, moving in a path of radius r, is proportional to the net torque acting on the mass.

$$\tau = mr^2 \alpha \qquad (8.5)$$

The equation for calculating moment of inertia of

(1) A **point mass** m, at a radial distance r, from a specified axis of rotation is shown in the figure at left.

(2) A **collection of discrete point masses,** each with its corresponding values of m and r is shown in the figure at right.

(3) **Extended rigid bodies** with a high degree of symmetry are shown in Table 8.1 of your textbook.

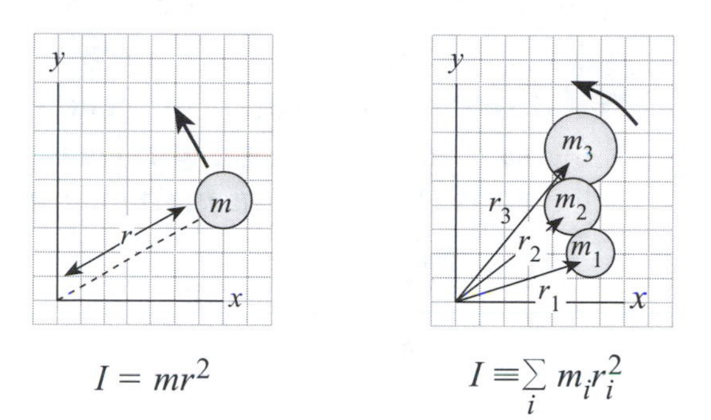

In all cases, the moment of inertia depends on the mass and distribution of the mass relative to the axis of rotation.

The **angular acceleration** of an extended object is proportional to the net torque acting on the object. *Equation 8.8 is the rotational analog of Newton's second law.*

$$\Sigma\tau = I\alpha \qquad (8.8)$$

Rotational kinetic energy is associated with a rigid body in rotational motion. *Note that this is not a new form of energy but is a convenient form for representing kinetic energy associated with rotational motion.*

$$KE_r \equiv \tfrac{1}{2} I\omega^2 \qquad (8.10)$$

When only conservative forces act on a system, the total mechanical energy of the system (sum of the gravitational, elastic potential, rotational kinetic, and translational kinetic energies) is conserved.

$$(KE_t + KE_r + PE)_i = (KE_t + KE_r + PE)_f \qquad (8.11)$$

When nonconservative forces act on a system, the total mechanical energy of the system changes by an amount equal to the work done by the nonconservative forces.

$$W_{nc} = \Delta KE_t + \Delta KE_r + \Delta PE \tag{8.12}$$

The **angular momentum** is a vector quantity associated with rotational motion.

$$L \equiv I\omega \tag{8.13}$$

The **magnitude** of angular momentum is defined by Equation 8.13. It must be determined relative to a specified axis due to its dependence on the moment of inertia.

The **direction** of angular momentum is the same as the direction of angular velocity as illustrated in the figure at right.

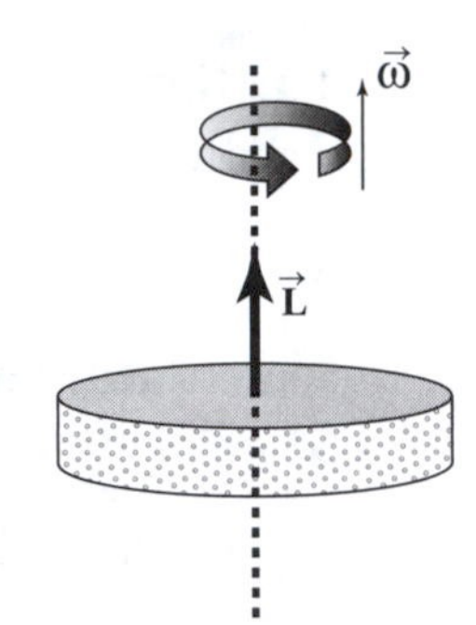

The **net external torque** acting on an object equals the time rate of change of its angular momentum. *Equation 8.14 is the rotational analog of Newton's second law for transitional motion.*

$$\Sigma\tau = \frac{\Delta L}{\Delta t} \tag{8.14}$$

The **law of conservation of angular momentum** states that if the resultant external torque acting on a system is zero, the total angular momentum will remain constant in both magnitude and direction.

$$L_i = L_f \quad \text{if} \quad \Sigma\tau = 0 \tag{8.15}$$

$$I_i\omega_i = I_f\omega_f \quad \text{if} \quad \Sigma\tau = 0 \tag{8.16}$$

SUGGESTIONS, SKILLS, AND STRATEGIES

PROBLEM-SOLVING STRATEGY FOR OBJECTS IN EQUILIBRIUM

1. Draw a simple, neat diagram of the system that is large enough to show all the forces clearly.

2. Isolate the object of interest being analyzed. Draw a free-body diagram for this object showing all external forces acting on the object. For systems containing more than one object, draw separate diagrams for each object. Do not include forces that the object exerts on its surroundings.

3. Establish convenient coordinate axes for each body and find the components of the forces along these axes. Now apply the first condition of equilibrium for each object under consideration; namely, that the net force on the object in the x- and y-directions must be zero.

4. Choose a convenient origin (axis of rotation) for calculating the net torque on the object. Now apply the second condition of equilibrium that says that the net torque on the object about any origin must be zero. Remember that the choice of the origin for the torque equation is arbitrary; therefore, choose an origin that will simplify your calculation as much as possible. Note that a force that acts along a line passing through the point chosen as the axis of rotation gives zero contribution to the torque because the lever arm is zero in this case.

5. The first and second conditions for equilibrium will give a set of simultaneous equations with several unknowns. To complete your solution, all that is left is to solve for the unknowns in terms of the known quantities.

PROBLEM-SOLVING STRATEGY FOR ROTATIONAL MOTION

The following facts and procedures should be kept in mind when solving rotational motion problems.

1. Problems involving the equation $\Sigma\vec{\boldsymbol{\tau}} = \mathbf{I}\vec{\boldsymbol{\alpha}}$ are very similar to those encountered in Newton's second law problems, $\Sigma\vec{\mathbf{F}} = m\vec{\mathbf{a}}$. Note the correspondences between linear and rotational quantities in that $\vec{\mathbf{F}}$ is replaced by $\vec{\boldsymbol{\tau}}$, m by I, and $\vec{\mathbf{a}}$ by $\vec{\boldsymbol{\alpha}}$.

2. Other analogues between rotational quantities and linear quantities include the replacement of x by θ and v by ω. Recall that each linear quantity (x, v, and a) equals the product of the radius and the corresponding angular quantity (θ, ω, and α). These are helpful as memory devices for such rotational motion quantities as rotational kinetic energy, $KE_r = \frac{1}{2}I\omega^2$, and angular momentum, $L = I\omega$.

3. With the analogues mentioned in step 2, conservation of energy techniques remain the same as those examined in Chapter 5, except now an additional term representing rotational kinetic energy must be included in the expression for the conservation of energy. See Equation 8.11.

4. Likewise, the techniques for solving conservation of angular momentum problems are essentially the same as those used in solving conservation of linear momentum problems, except in this case you will be equating total initial and final angular momentum values: $I_i\omega_i = I_f\omega_f$.

REVIEW CHECKLIST

- There are two necessary conditions for equilibrium of a rigid body: $\Sigma F = 0$ and $\Sigma\tau = 0$. Torques that cause counterclockwise rotations are positive and those causing clockwise rotations are negative.

- The torque associated with a force has a magnitude equal to the force times the lever arm. The lever arm is the perpendicular distance from the axis of rotation to a line drawn along the direction of the force. Also, the net torque on a rigid body about some axis is proportional to the angular acceleration; i.e., $\Sigma\tau = I\alpha$, where I is the moment of inertia about the axis for which the net torque is evaluated.

- The work-energy theorem can be applied to a rotating rigid body. That is, the net work done on a rigid body rotating about a fixed axis equals the change in its rotational kinetic energy. The law of conservation of mechanical energy can be used in the solution of problems involving rotating rigid bodies.

- The time rate of change of the angular momentum of a rigid body rotating about an axis is proportional to the net torque acting about the axis of rotation. This is the rotational analog of Newton's second law.

SOLUTIONS TO SELECTED END-OF-CHAPTER PROBLEMS

7. The arm in Figure P8.7 weighs 41.5 N. The force of gravity acting on the arm acts through point A. Determine the magnitudes of the tension force $\vec{\mathbf{F}}_t$ in the deltoid muscle and the force $\vec{\mathbf{F}}_s$ exerted by the shoulder on the humerus (upper-arm bone) to hold the arm in the position shown.

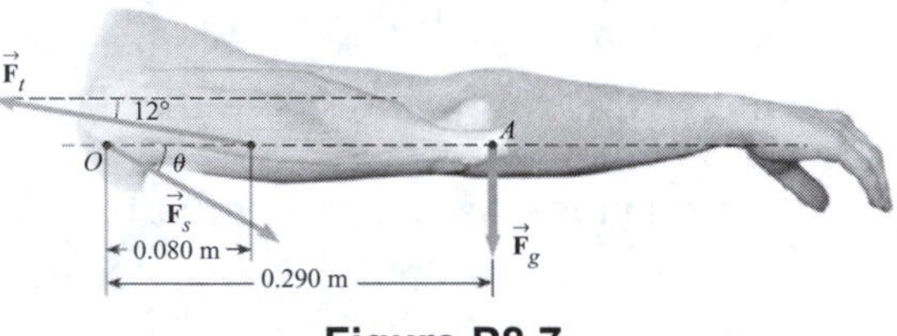

Figure P8.7

Solution

For the easiest solution of this problem, one should resolve the unknown forces $\vec{\mathbf{F}}_t$ and $\vec{\mathbf{F}}_s$ into their horizontal and vertical components as shown in the free-body diagram at the right. Then, it is observed that the lines of action of many of the unknown components pass through point O, meaning that these components have zero torque about a pivot at point O. Thus, choosing point O as the pivot when applying the second condition for equilibrium, $\Sigma\vec{\tau} = 0$, will lead to a simple equation to solve:

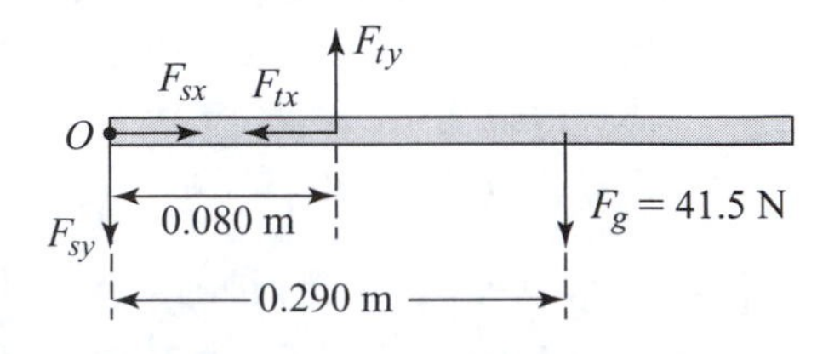

$$\Sigma\vec{\tau}\Big)_O = 0 \quad \Rightarrow \quad F_{ty}(0.080 \text{ m}) - (41.5 \text{ N})(0.290 \text{ m}) = 0$$

and since $F_{ty} = F_t \sin 12.0°$, the magnitude of the tension force is

$$F_t = \frac{(41.5 \text{ N})(0.290 \text{ m})}{(0.080 \text{ m})\sin 12.0°} = 724 \text{ N} \qquad \lozenge$$

Then

$$\Sigma F_x = 0 \quad \Rightarrow \quad F_{sx} = F_{tx} = F_t \cos 12.0° = (724 \text{ N})\cos 12.0° = 708 \text{ N}$$

and

$$\Sigma F_y = 0 \quad \Rightarrow \quad F_{sy} = F_{ty} - F_g = (724 \text{ N})\sin 12.0° - 41.5 \text{ N} = 109 \text{ N}$$

so the magnitude of the shoulder force is

$$F_s = \sqrt{F_{sx}^2 + F_{sy}^2} = \sqrt{(708 \text{ N})^2 + (109 \text{ N})^2} = 716 \text{ N} \qquad \Diamond$$

---

11. Find the x- and y-coordinates of the center of gravity of a 4.00-ft by 8.00-ft uniform sheet of plywood with the upper right quadrant removed as shown in Figure P8.11. [*Hint:* The mass of any segment of the plywood sheet is proportional to the area of that segment.]

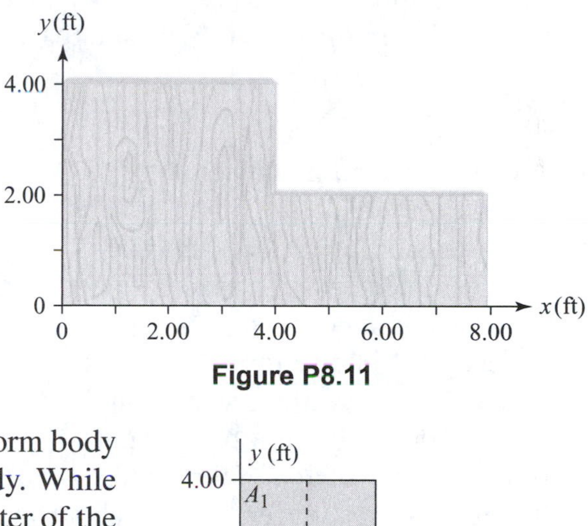

Figure P8.11

Solution

The location of the center of gravity of a uniform body is located at the geometric center of that body. While we cannot easily recognize the geometric center of the irregularly shaped sheet of plywood in Figure P8.11, we can think of this sheet as consisting of two areas A_1 and A_2 as shown in the sketch at the right.

The geometric center (and hence center of gravity) of each of these areas can be found by inspection. The center of gravity of A_1 is located at $(x_1 = 2.00 \text{ ft}, y_1 = 2.00 \text{ ft})$ and that of A_2 is at $(x_2 = 6.00 \text{ ft}, y_2 = 1.00 \text{ ft})$. The mass of each portion of plywood is $m_1 = \sigma A_1$ and $m_2 = \sigma A_2$, where σ is the constant mass per unit area of the uniform sheet. Since the mass of each portion may be considered to be concentrated at its center of gravity, the coordinates of the center of gravity of the irregular plywood sheet are

$$x_{cg} = \frac{\Sigma m_i x_i}{\Sigma m_i} = \frac{(\not{\sigma} A_1)x_1 + (\not{\sigma} A_2)x_2}{\not{\sigma} A_1 + \not{\sigma} A_2} = \frac{(16.0 \text{ ft}^2)(2.00 \text{ ft}) + (8.00 \text{ ft}^2)(6.00 \text{ ft})}{16.0 \text{ ft}^2 + 8.00 \text{ ft}^2} = 3.33 \text{ ft} \qquad \Diamond$$

and

$$y_{cg} = \frac{\Sigma m_i y_i}{\Sigma m_i} = \frac{(\not{\sigma} A_1)y_1 + (\not{\sigma} A_2)y_2}{\not{\sigma} A_1 + \not{\sigma} A_2} = \frac{(16.0 \text{ ft}^2)(2.00 \text{ ft}) + (8.00 \text{ ft}^2)(1.00 \text{ ft})}{16.0 \text{ ft}^2 + 8.00 \text{ ft}^2} = 1.67 \text{ ft} \qquad \Diamond$$

---

17. A person bending forward to lift a load "with his back" (Fig. P8.17) rather than "with his knees" can be injured by large forces exerted on the muscles and vertebrae. The spine pivots mainly at the fifth lumbar vertebra, with the principal supporting force provided by the erector spinalis muscle in the back. To see the magnitude of the forces involved, and to understand why back problems are common among humans, consider the model shown in Figure P8.17b of a person bending forward to lift a 200-N object. The spine and upper body are represented as a uniform horizontal rod of weight 350 N, pivoted at the base of the spine. The erector spinalis muscle, attached at a point two-thirds of the way up the spine, maintains the position of the back. The angle between the spine and this muscle is 12.0°. Find (a) the tension in the back muscle and (b) the compressional force in the spine.

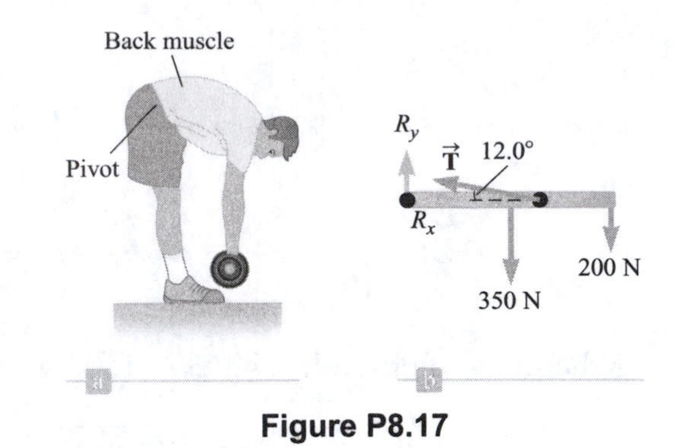

Figure P8.17

Solution

(a) Consider the force diagram of the spine given at the right, where the tension in the back muscle is represented by its horizontal and vertical components. Point O is a very good choice for a pivot point when applying the second condition for equilibrium to the spine. The lines of action of three unknown forces pass through this point, and hence, those forces will have zero torque about this point. Using this pivot point, the resulting torque equation is

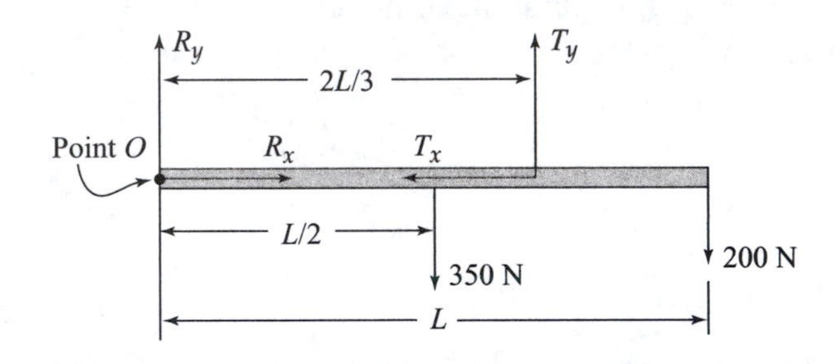

$$\Sigma\tau_O = 0 + 0 + 0 - (350\text{ N})\left(\frac{\not{L}}{2}\right) + T_y\left(\frac{2\not{L}}{3}\right) - (200\text{ N})\not{L} = 0$$

which reduces to

$$T_y = \frac{3}{2}\left(\frac{350\text{ N}}{2} + 200\text{ N}\right) = 563\text{ N}$$

From Figure P8.17b, we see that $T_y = T\sin 12.0°$, so the tension in the back muscle is

$$T = \frac{T_y}{\sin 12.0°} = \frac{563\text{ N}}{\sin 12.0°} = 2.71\times 10^3\text{ N} = 2.71\text{ kN} \qquad \Diamond$$

(b) Applying the first condition of equilibrium to the horizontal forces acting on the spine gives $\Sigma F_x = R_x - T_x = 0$, so the compression force in the spine is

$$R_x = T_x = T\cos 12.0° = (2.71\text{ kN})\cos 12.0° = 2.65\text{ kN} \qquad \Diamond$$

30. One end of a uniform 4.0-m-long rod of weight w is supported by a cable at an angle of $\theta = 37°$ with the rod. The other end rests against a wall, where it is held by friction. (See Fig. P8.30.) The coefficient of static friction between the wall and the rod is $\mu_s = 0.50$. Determine the minimum distance x from point A at which an additional weight w (the same as the weight of the rod) can be hung without causing the rod to slip at point A.

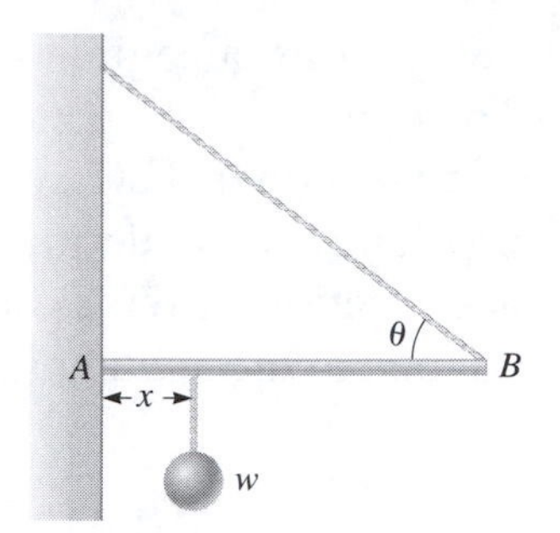

Figure P8.30

Solution

(a) In the force diagram of the rod, the tension in the cable is replaced by its components $T_x = T\cos 37°$ and $T_y = T\sin 37°$. The wall exerts a normal force n and an upward static friction force f_s on the left end of the rod.

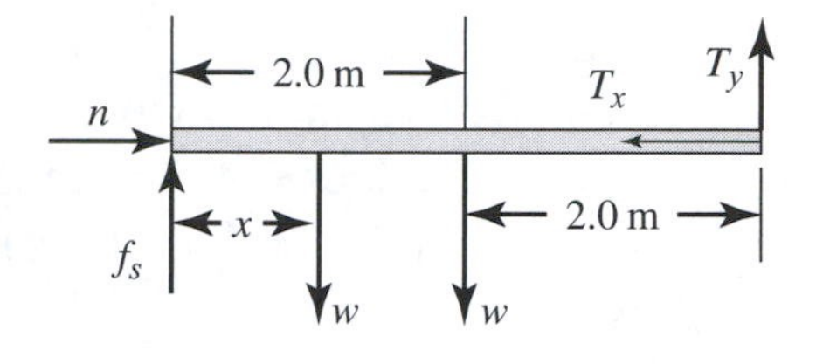

While the rod remains in equilibrium, we may require that $\Sigma F_x = 0$ and $\Sigma F_y = 0$. These give

$$\Sigma F_x = n - T\cos 37° = 0 \qquad \text{or} \qquad n = 0.80T \qquad [1]$$

and $$\Sigma F_y = f_s + T\sin 37° - 2w = 0 \qquad \text{or} \qquad f_s + 0.60T = 2w \qquad [2]$$

When $x = x_{min}$ and the left end of the rod is about to slip on the wall, $f_s = (f_s)_{max} = \mu_s n$. Using Equation [1], this becomes $f_s = (0.50)(0.80T) = 0.40T$. Substituting this result into Equation [2] yields $(0.40 + 0.60)T = 2w$, or when $x = x_{min}$ and the rod is ready to slip, we have $T = 2w$. We now make use of this result and the second condition of equilibrium to solve for the minimum value the distance x may be without the rod slipping.

Requiring that the sum of the torques be zero about the left end of the rod gives

$$\Sigma\tau_{\substack{\text{left}\\\text{end}}} = 0 + 0 - w\cdot x - w(2.0\text{ m}) + 0 + (T\sin 37°)(4.0\text{ m}) = 0$$

Making use of our knowledge that $T = 2w$ when $x = x_{min}$, the torque equation becomes $-\cancel{w}\cdot x_{min} - \cancel{w}(2.0\text{ m}) + 0 + (2\cancel{w})(0.60)(4.0\text{ m}) = 0$, and yields

$$x_{min} = 2.8\text{ m}$$ ◊

35. A rope of negligible mass is wrapped around a 225-kg solid cylinder of radius 0.400 m. The cylinder is suspended several meters off the ground with its axis oriented horizontally, and turns on that axis without friction. (a) If a 75.0-kg man takes hold of the free end of the rope and falls under the force of gravity, what is his acceleration? (b) What is the angular acceleration of the cylinder? (c) If the mass of the rope were not neglected, what would happen to the angular acceleration of the cylinder as the man falls?

Solution

(a) The sketch at the right includes force diagrams of the cylinder and the man. We shall assume the accelerations are positive in the directions shown. This means that clockwise torques are considered positive and downward is the positive direction for the man. If the rope does not slip on the cylinder, the angular acceleration of the cylinder and the linear acceleration of the man are related by $a = r\alpha$.

Applying the translational form of Newton's second law to the man gives

$$\Sigma F_y = mg - T = ma \quad \text{or} \quad T = m(g-a) \qquad \textbf{[1]}$$

Next, we apply the rotational form of Newton's second law to the cylinder, computing torques about the axis of the cylinder. This gives $\Sigma\tau = Tr = I\alpha$. Using the facts that $I = \frac{1}{2}Mr^2$ for a solid cylinder and $\alpha = a/r$ if the rope does not slip, this becomes

$$Tr = \frac{1}{2}Mr^2\left(\frac{a}{r}\right) \quad \text{or} \quad T = \frac{1}{2}Ma \qquad \textbf{[2]}$$

Substituting Equation [2] into Equation [1] and simplifying yields

$$a = \frac{mg}{m + (M/2)}$$

With $M = 225$ kg and $m = 75.0$ kg, the downward acceleration of the man is

$$a = \frac{(75.0\ \text{kg})(9.80\ \text{m/s}^2)}{75.0\ \text{kg} + (225\ \text{kg}/2)} = 3.92\ \text{m/s}^2 \qquad \Diamond$$

(b) The angular acceleration of the cylinder, with $r = 0.400$ m, is then

$$\alpha = \frac{a}{r} = \frac{3.92\ \text{m/s}^2}{0.400\ \text{m}} = 9.80\ \text{rad/s}^2 \qquad \Diamond$$

(c) If the mass of the rope was not negligible, two factors would combine to produce an increasing angular acceleration of the cylinder. First, as rope is unwound from the cylinder, the remaining mass (and hence, moment of inertia) of the cylinder would be decreasing. Second, the mass of the rope that has already left the cylinder would combine with the mass of the man to increase the tension in the rope at the point where it leaves the cylinder. This would produce an increasing torque acting on the cylinder. Both of these factors would increase the angular acceleration of the cylinder (and hence, the linear acceleration of the man).

---

41. An airliner lands with a speed of 50.0 m/s. Each wheel of the plane has a radius of 1.25 m and a moment of inertia of $110\ \text{kg}\cdot\text{m}^2$. At touchdown, the wheels begin to spin under the action of friction. Each wheel supports a weight of 1.40×10^4 N, and the wheels attain their angular speed in 0.480 s while rolling without slipping. What is the coefficient of kinetic friction between the wheels and the runway? Assume that the speed of the plane is constant.

Solution

(a) Consider the force diagram of the wheel given at the right and recognize that the wheel has zero vertical acceleration. Thus, $\Sigma F_y = n - F_y = 0$ and the normal force is $n = F_y = 1.40\times10^4$ N. Since the wheel rolls without slipping, the final angular velocity of the wheel is related to the translational speed of the axle by $v = r\omega_f$, giving

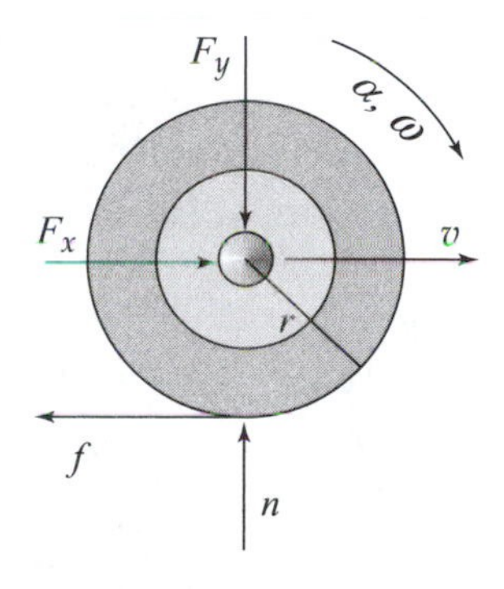

$$\omega_f = \frac{v}{r} = \frac{50.0\ \text{m/s}}{1.25\ \text{m}} = 40.0\ \text{rad/s}$$

The initial angular velocity of the wheel was $\omega_i = 0$, and the change in angular velocity occurred in a time interval of $\Delta t = 0.480$ s. Therefore, the average angular acceleration was

$$\alpha = \frac{\Delta\omega}{\Delta t} = \frac{40.0\ \text{rad/s} - 0}{0.480\ \text{s}} = 83.3\ \text{rad/s}^2$$

Computing torques about the center of the wheel, taking clockwise as positive, the rotational form of Newton's second law gives $\Sigma\tau = f\cdot r = I\alpha$. Thus, the average friction force acting on the wheel during the acceleration period must have been

$$f = \frac{I\alpha}{r} = \frac{(110\ \text{kg}\cdot\text{m}^2)(83.3\ \text{rad/s}^2)}{1.25\ \text{m}} = 7.33\times10^3\ \text{N}$$

and the coefficient of kinetic friction is

$$\mu_k = \frac{f}{n} = \frac{7.33\times10^3\ \text{N}}{1.40\times10^4\ \text{N}} = 0.524$$ ◊

---

47. A solid, uniform disk of radius 0.250 m and mass 55.0 kg rolls down a ramp of length 4.50 m that makes an angle of 15.0° with the horizontal. The disk starts from rest from the top of the ramp. Find (a) the speed of the disk's center of mass when it reaches the bottom of the ramp and (b) the angular speed of the disk at the bottom of the ramp.

Solution

(a) As the disk rolls down the ramp, two nonconservative forces (a normal force and a friction force) are exerted on the disk by the ramp. The normal force is perpendicular to the motion of the disk, and hence, does no work on the disk. Assuming the disk rolls without slipping, the point on the disk in contact with the ramp at each instant is momentarily at rest, and the friction force does no work on the disk. Thus, the total mechanical energy of the disk is constant as it rolls from the top to the bottom of the ramp, or

$$\left(KE_t + KE_r + PE_g\right)_f = \left(KE_t + KE_r + PE_g\right)_i$$

Taking $y = 0$ (and hence $PE_g = 0$) at the bottom of the ramp, this becomes

$$\frac{1}{2}mv_f^2 + \frac{1}{2}I\omega_f^2 + 0 = 0 + 0 + mgy_i$$

When an object rolls without slipping, $v = r\omega$, so $\omega = v/r$. Also, $I = \frac{1}{2}mr^2$ for a uniform, solid disk. Our conservation of energy equation then becomes

$$\frac{1}{2}\cancel{m}v_f^2 + \frac{1}{2}\left(\frac{\cancel{m}\,\cancel{r^2}}{2}\right)\left(\frac{v_f^2}{\cancel{r^2}}\right) = \cancel{m}gy_i \quad \text{or} \quad \frac{3}{4}v_f^2 = gy_i$$

and $$v_f = \sqrt{\frac{4gy_i}{3}} = \sqrt{\frac{4\left(9.80\ \text{m/s}^2\right)(4.50\ \text{m}\cdot\sin 15.0°)}{3}} = 3.90\ \text{m/s} \qquad \Diamond$$

(b) The angular speed of the disk when it reaches the bottom of the ramp is then

$$\omega_f = \frac{v_f}{r} = \frac{3.90\ \text{m/s}}{0.250\ \text{m}} = 15.6\ \text{rad/s} \qquad \Diamond$$

53. A giant swing at an amusement park consists of a 365-kg uniform arm 10.0 m long, with two seats of negligible mass connected at the lower end of the arm (Fig. P8.53). (a) How far from the upper end is the center of mass of the arm? (b) The gravitational potential energy of the arm is the same as if all its mass were concentrated at the center of mass. If the arm is raised through a 45.0° angle, find the gravitational potential energy, where the zero level is taken to be 10.0 m below the axis. (c) The arm drops from rest from the position described in part (b). Find the gravitational potential energy of the system when it reaches the vertical orientation. (d) Find the speed of the seats at the bottom of the swing.

Solution

(a) The arm consists of a uniform rod 10.0 m long and the mass of the seats at the lower end is negligible. Thus, the center of gravity of the arm is located at its geometric center, 5.00 m from either end. ◊

From the sketch at the right, the height of the center of gravity above ground level is

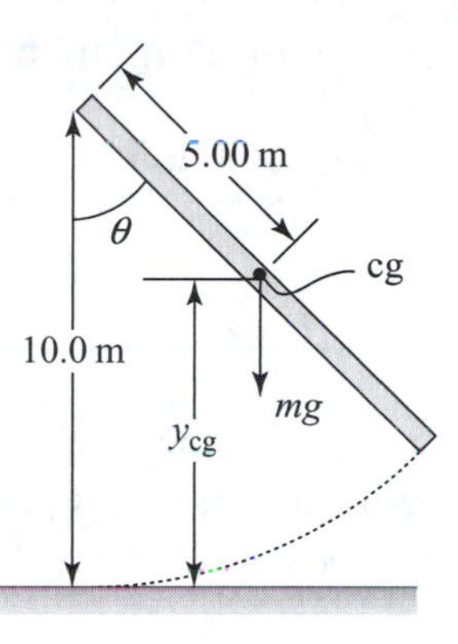

$$y_{cg} = 10.0 \text{ m} - (5.00 \text{ m})\cos\theta$$

(b) When $\theta = 45.0°$, $y_{cg} = 10.0 \text{ m} - (5.00 \text{ m})\cos 45.0°$

and $PE_g = mgy_{cg}$ gives the gravitational potential energy as

$$PE_g = (365 \text{ kg})(9.80 \text{ m/s}^2)[10.0 \text{ m} - (5.00 \text{ m})\cos 45.0°] = 2.31\times 10^4 \text{ J}$$ ◊

(c) In the vertical orientation, $\theta = 0°$ and $\cos\theta = 1$, giving $y_{cg} = 5.00$ m.

Then, $PE_g = mgy_{cg} = (365 \text{ kg})(9.80 \text{ m/s}^2)(5.00 \text{ m}) = 1.79\times 10^4 \text{ J}$ ◊

(d) When the arm rotates about an axis, perpendicular to its length and through the upper end, the moment of inertia is $I_{end} = \frac{1}{3}mL^2$. We then apply conservation of mechanical energy as the arm starts from rest in the 45.0° orientation and obtain $\frac{1}{2}I_{end}\omega_f^2 + mg(y_{cg})_f = 0 + mg(y_{cg})_i$, yielding

$$\omega_f = \sqrt{\frac{2mg\left[(y_{cg})_i - (y_{cg})_f\right]}{mL^2/3}} = \frac{\sqrt{6g\left[(y_{cg})_i - (y_{cg})_f\right]}}{L}$$

The translational speed of the seats, following a circular path of radius $r = L$, when the arm reaches the vertical orientation is

$$v = r\omega_f = L\omega_f = \sqrt{6g\left[(y_{cg})_i - (y_{cg})_f\right]}$$

or $$v = \sqrt{6(9.80 \text{ m/s}^2)([10.0 - 5.00\cos 45.0°] \text{ m} - 5.00 \text{ m})} = 9.28 \text{ m/s}$$ ◊

61. A solid, horizontal cylinder of mass 10.0 kg and radius 1.00 m rotates with an angular speed of 7.00 rad/s about a fixed vertical axis through its center. A 0.250-kg piece of putty is dropped vertically onto the cylinder at a point 0.900 m from the center of rotation, and sticks to the cylinder. Determine the final angular speed of the system.

Solution

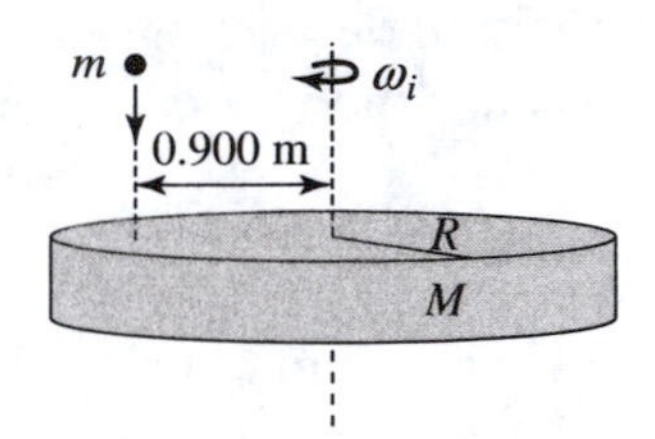

Before the putty sticks to the cylinder, the moment of inertia of this solid cylinder about an axis through its center is

$$I_{\text{cylinder}} = \frac{1}{2}MR^2 = \frac{1}{2}(10.0\ \text{kg})(1.00\ \text{m})^2 = 5.00\ \text{kg}\cdot\text{m}^2$$

Just before its impact with the cylinder, the putty is moving parallel to the rotation axis, with no movement around that axis. Therefore, the putty has zero angular momentum about the rotation axis before impact.

The total angular momentum about the rotation axis before the putty sticks to the cylinder is then

$$L_i = L_{\text{cylinder}} + L_{\text{putty}} = I_{\text{cylinder}}\omega_i + 0 = \left(5.00\ \text{kg}\cdot\text{m}^2\right)\left(7.00\ \text{rad/s}\right) = 35.0\ \text{kg}\cdot\text{m}^2/\text{s}$$

After the putty strikes and sticks to the cylinder, both objects rotate about the rotation axis with the same angular speed ω_f. The total moment of inertia of the rotating system about the axis is now

$$I_f = I_{\text{cylinder}} + I_{\text{putty}} = 5.00\ \text{kg}\cdot\text{m}^2 + mr^2 = 5.00\ \text{kg}\cdot\text{m}^2 + (0.250\ \text{m})(0.900\ \text{m})^2$$

or $\quad I_f = 5.20\ \text{kg}\cdot\text{m}^2$

Since no external forces exert a torque about the rotation axis on the cylinder-putty system, the angular momentum of that system is conserved. Thus,

$$L_f = I_f\omega_f = L_i \qquad \text{and} \qquad \omega_f = \frac{L_i}{I_f}$$

or the final angular speed of the system is $\quad \omega_f = \dfrac{35.0\ \text{kg}\cdot\text{m}^2/s}{5.20\ \text{kg}\cdot\text{m}^2} = 6.73\ \text{rad/s} \quad \lozenge$

71. A uniform ladder of length L and weight w is leaning against a vertical wall. The coefficient of static friction between the ladder and the floor is the same as that between the ladder and the wall. If this coefficient of static friction is $\mu_s = 0.500$, determine the smallest angle the ladder can make with the floor without slipping.

Solution

If the ladder is to slip, it must slip at the wall and at the floor simultaneously. Thus, at the critical angle of inclination θ, the static friction forces at the floor and wall must be at their maximum values, $(f_s)_{max} = \mu_s n$, as shown in the force diagram of the ladder at the right.

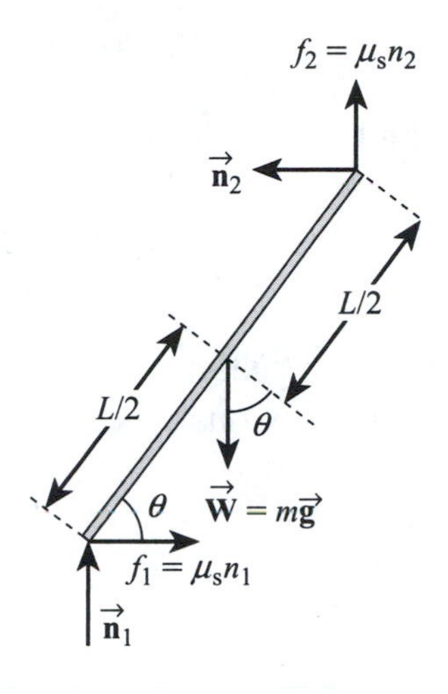

From the first condition of equilibrium,

$$\Sigma F_x = \mu_s n_1 - n_2 = 0$$

or

$$n_2 = \mu_s n_1 \qquad [1]$$

and

$$\Sigma F_y = n_1 + \mu_s n_2 - w = 0$$

or

$$n_1 + \mu_s(\mu_s n_1) = w$$

yielding

$$n_1 = \frac{w}{1+\mu_s^2} = \frac{w}{1+(0.500)^2}$$

or

$$n_1 = 0.800\,w \qquad [2]$$

Substitution of Equation [2] into [1] gives

$$n_2 = (0.500)(0.800\,w) = 0.400\,w \qquad [3]$$

For the second condition of equilibrium, we consider the torques about an axis perpendicular to the page and passing through the lower end of the ladder. This gives

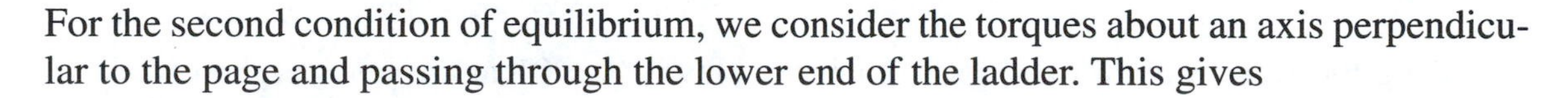

$$\Sigma\tau_{\substack{\text{lower}\\\text{end}}} = -w\left(\frac{\cancel{L}}{2}\cos\theta\right) + n_2(\cancel{L}\sin\theta) + \mu_s n_2(\cancel{L}\cos\theta) = 0$$

Using Equation [3], this becomes

$$\frac{-\cancel{w}}{2}\cos\theta + 0.400\,\cancel{w}\,\sin\theta + (0.500)\left(0.400\,\cancel{w}\right)\cos\theta = 0$$

or

$$0.400\sin\theta = (0.500 - 0.200)\cos\theta$$

and

$$\tan\theta = \frac{\sin\theta}{\cos\theta} = \frac{0.300}{0.400} = 0.750$$

The critical angle where the ladder is on the verge of slipping at the floor and at the wall simultaneously is

$$\theta = \tan^{-1}(0.750) = 36.9^\circ \qquad \lozenge$$

---

83. A *war-wolf*, or *trebuchet*, is a device used during the Middle Ages to throw rocks at castles and now sometimes used to fling pumpkins and pianos. A simple trebuchet is shown in Figure P8.83. Model it as a stiff rod of negligible mass 3.00 m long and joining particles of mass $m_1 = 0.120$ kg and $m_2 = 60.0$ kg at its ends. It can turn on a frictionless horizontal axle perpendicular to the rod and 14.0 cm from the particle of larger mass. The rod is released from rest in a horizontal orientation. Find the maximum speed that the object of smaller mass attains.

Figure P8.83

Solution

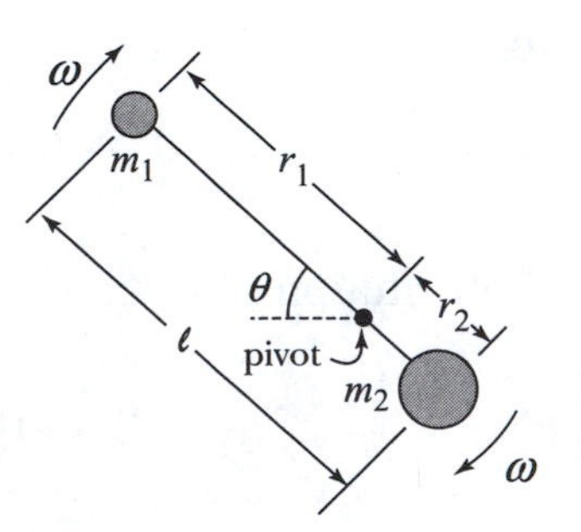

The sketch at the right shows the model of the trebuchet when it is inclined at angle θ to the horizontal. Its overall length is $\ell = 3.00$ m. Mass m_2 is distance $r_2 = 0.140$ m from the pivot, so mass m_1 is at $r_1 = \ell - r_2 = 2.86$ m. Neglecting any mass of the rod, the moment of inertia about the axis perpendicular to the rod and through the pivot is $I = m_1 r_1^2 + m_2 r_2^2$, or

$$I = (0.120\text{ kg})(2.86\text{ m})^2 + (60.0\text{ kg})(0.140\text{ m})^2 = 2.16\text{ kg}\cdot\text{m}^2$$

Choosing the zero gravitational potential energy level $(y = 0)$ at the level of the pivot, the total gravitational potential energy of the system when the rod is tilted at angle θ to the horizontal is

$$PE_g = m_1 g y_1 + m_2 g y_2 = m_1 g r_1 \sin\theta - m_2 g r_2 \sin\theta = \left(m_1 r_1 - m_2 r_2\right) g \sin\theta$$

The kinetic energy associated with the pure rotational motion of the trebuchet is $KE_r = \frac{1}{2}I\omega^2$. The smaller mass m_1 has maximum speed when the larger mass m_2 is at its lowest point and the rod is vertical. As the trebuchet rotates about the pivot, no non-conservative forces do work on it, so the total mechanical energy is conserved. Applying $\left(KE_r + PE_g\right)_f = \left(KE_r + PE_g\right)_i$ from when the trebuchet starts from rest ($\omega_i = 0$ and $\theta_i = 0$) until it reaches the vertical ($\theta = 90°$), we have

$$\frac{1}{2}I\omega_f^2 + \left(m_1 r_1 - m_2 r_2\right)g\sin 90° = 0 + 0$$

which yields $\omega_f = \sqrt{2g\left(m_2 r_2 - m_1 r_1\right)/I}$. The maximum translational speed of mass m_1 is then $v_f = r_1\omega_f$, or

$$v_f = (2.86\text{ m})\sqrt{\frac{2\left(9.80\text{ m/s}^2\right)\left[(60.0\text{ kg})(0.140\text{ m}) - (0.120\text{ kg})(2.86\text{ m})\right]}{2.16\text{ kg}\cdot\text{m}^2}} = 24.5\text{ m/s} \quad \lozenge$$

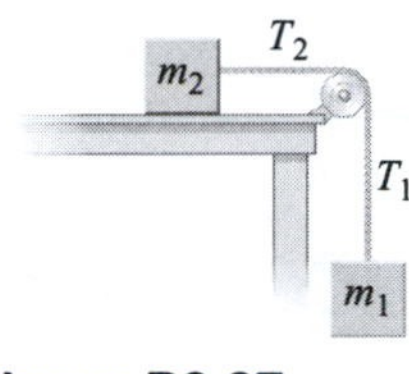

Figure P8.87

87. An object of mass $m_1 = 4.00$ kg is connected by a light cord to an object of mass $m_2 = 3.00$ kg on a frictionless surface (Fig. P8.87). The pulley rotates about a frictionless axle and has a moment of inertia of $0.500\text{ kg}\cdot\text{m}^2$ and a radius of 0.300 m. Assuming that the cord does not slip on the pulley, find (a) the acceleration of the two masses and (b) the tensions T_1 and T_2.

Solution

(a) In the diagrams given at the right of the two masses and the pulley, notice the directions and signs given the accelerations. We will take counter-clockwise rotations (and torques) as positive, and upward as the positive vertical direction. Observe that the two masses have the same magnitude acceleration and this is related to the angular acceleration of the pulley by $\alpha = -(a/r)$ [negative because the pulley accelerates clockwise]. Applying Newton's second law to the motion of each object gives:

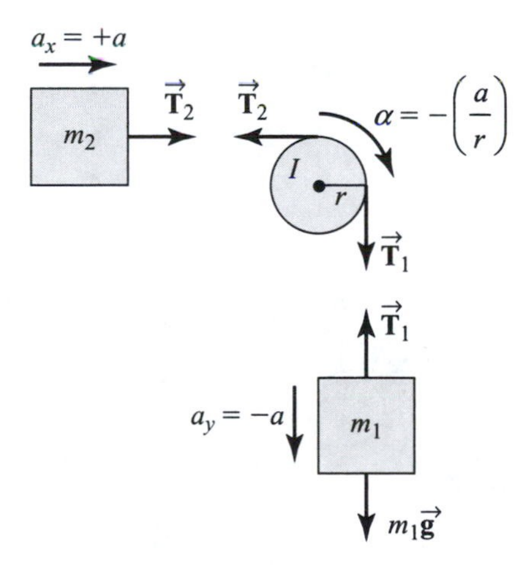

For m_1: $\quad \Sigma F_y = m_1 a_y \quad \Rightarrow \quad T_1 - m_1 g = m_1(-a) \quad$ or $\quad T_1 = m_1\left(g - a\right)$ **[1]**

For m_2: $\quad \Sigma F_x = m_2 a_x \quad \Rightarrow \quad T_2 = m_2 a$ **[2]**

For the pulley: $\Sigma\tau = I\alpha \quad \Rightarrow \quad +T_2 r - T_1 r = I\left(-a/r\right) \quad$ or $\quad T_1 - T_2 = \left(I/r^2\right)a$ **[3]**

Substituting Equations [1] and [2] into Equation [3] yields

$$m_1 g-(m_1+m_2)a=(I/r^2)a \qquad \text{or} \qquad a=\frac{m_1 g}{m_1+m_2+I/r^2}$$

The magnitude of the translational acceleration of the two masses is then

$$a=\frac{(4.00\ \text{kg})(9.80\ \text{m/s}^2)}{4.00\ \text{kg}+3.00\ \text{kg}+(0.500\ \text{kg}\cdot\text{m}^2)/(0.300\ \text{m})^2}=3.12\ \text{m/s}^2$$ ◊

(b) Substituting the result from part (a) into Equations [1] and [2] gives:

$$T_1=(4.00\ \text{kg})(9.80\ \text{m/s}^2-3.12\ \text{m/s}^2)=26.7\ \text{N}$$ ◊

and $T_2=(3.00\ \text{kg})(3.12\ \text{m/s}^2)=9.36\ \text{N}$ ◊

9

Solids and Fluids

NOTES ON SELECTED CHAPTER SECTIONS

9.1 States of Matter

Normal matter is generally classified as being in one of three states: solid, liquid, or gas. A fourth state consisting of an ionized gas is called a plasma.

In a **crystalline solid**, the atoms are arranged in an ordered periodic structure; while in an **amorphous solid** (i.e., glass), the atoms are present in a disordered fashion.

In the **liquid state**, thermal agitation is greater than in the solid state, the molecular forces are weaker, and molecules wander throughout the liquid in a random fashion.

The molecules of a **gas** are in constant random motion and exert weak forces on each other. The distances separating molecules are large compared to the dimensions of the molecules.

At very high temperature some gases become a **plasma** consisting of positive and negative ions which interact via electric and magnetic forces.

9.2 Density and Pressure

The **density**, ρ, of a substance of uniform composition is defined as its mass per unit volume and has units of kilograms per cubic meter (kg/m^3) in the SI system.

The **specific gravity** of a substance is a dimensionless quantity that is the ratio of the density of the substance to the density of water.

The **pressure**, P, in a fluid is the force per unit area that the fluid exerts on an object immersed in the fluid.

9.3 The Deformation of Solids

The **elastic properties of solids** are described in terms of stress and strain. Stress is a quantity that is related to the force causing a deformation; strain is a measure of the degree of deformation. It is found that, for sufficiently small stresses, stress is proportional to strain, and the constant of proportionality depends on the material being deformed and on the nature of the deformation. We call this proportionality constant the elastic modulus.

We shall consider three types of deformation and define an elastic modulus for each:

- **Young's modulus**, which measures the resistance of a solid to a change in its length
- **Shear modulus**, which measures the resistance to displacement of the planes of a solid sliding past each other
- **Bulk modulus**, which measures the resistance that solids or liquids offer to changes in their volume

9.4 Variation of Pressure with Depth

In a fluid at rest, all points at the same depth are at the same pressure. **Pascal's law** states that a change in pressure applied to an enclosed fluid is transmitted undiminished to every point in the fluid and the walls of the containing vessel. *The pressure, P, at a depth of h below the surface of a liquid with density ρ and open to the atmosphere is greater than atmospheric pressure by an amount ρgh.*

9.5 Pressure Measurements

The **absolute pressure** of a fluid is the sum of the **gauge pressure** and **atmospheric pressure**. The SI unit of pressure is the Pascal (Pa). Note that $1 \text{ Pa} \equiv 1 \text{ N/m}^2$.

9.6 Buoyant Forces and Archimedes's Principle

Any object partially or completely submerged in a fluid experiences a **buoyant force** equal in magnitude to the weight of the fluid displaced by the object and acting vertically upward through the point which was the center of gravity of the displaced fluid.

9.7 Fluids in Motion

Many features of fluid motion can be understood by considering the behavior of an ideal fluid, which satisfies the following conditions:

- **The fluid is nonviscous**; that is, there is no internal friction force between adjacent fluid layers.
- **The fluid is incompressible**, which means that its density is constant.
- **The fluid motion is steady**, meaning that the velocity, density, and pressure at each point in the fluid do not change in time.
- **The fluid moves without turbulence**. This implies that each element of the fluid has zero angular velocity about its center; that is, there can be no eddy currents present in the moving fluid.

Fluids that have the "ideal" properties stated above obey two important equations:

- The **equation of continuity** states that the flow rate through a pipe is constant (i.e., the product of the cross-sectional area of the pipe and the speed of the fluid is constant).
- **Bernoulli's equation** states that the sum of the pressure (P), kinetic energy per unit volume ($\rho v^2/2$), and the potential energy per unit volume (ρgh) has a constant value at all points along a streamline.

9.9 Surface Tension, Capillary Action, and Viscous Fluid Flow

The concept of **surface tension** can be thought of as the energy content of the fluid at its surface per unit surface area. In general, any equilibrium configuration of an object is one in which the energy is minimum. For a given volume, the spherical shape is the one that has the smallest surface area; therefore, a drop of water takes on a spherical shape. The surface tension of liquids decreases with increasing temperature.

Forces between like molecules, such as the forces between water molecules, are called **cohesive forces**, and forces between unlike molecules, such as those of glass on water, are **adhesive forces**. If a capillary tube is inserted into a fluid for which adhesive forces dominate over cohesive forces, the surrounding liquid will rise into the tube. If a capillary tube is inserted into a liquid in which cohesive forces dominate over adhesive forces, the level of the liquid in the capillary tube will be below the surface of the surrounding fluid.

Viscosity refers to the internal friction of a fluid. At sufficiently high velocities, fluid flow changes from simple streamline flow to turbulent flow. The onset of turbulence in a tube is determined by a factor called the **Reynolds number**, which is a function of the density of the fluid, the average speed of the fluid along the direction of flow, the diameter of the tube, and the viscosity of the fluid.

9.10 Transport Phenomena

The two fundamental processes involved in fluid transport resulting from concentration differences are called **diffusion** and **osmosis**. In a diffusion process, molecules move from a region where their concentration is high to a region where their concentration is lower. Diffusion occurs readily in air; the process also occurs in liquids and, to a lesser extent, in solids. Osmosis is defined as the movement of water from a region where its concentration is high, across a selectively permeable membrane, into a region where its concentration is lower. Osmosis is often described simply as the diffusion of water across a membrane.

EQUATIONS AND CONCEPTS

The **density** of a homogeneous substance is defined as its ratio of mass per unit volume. *Density is characteristic of a particular type of material and independent of the total quantity of material in the sample.*

$$\rho \equiv \frac{M}{V} \tag{9.1}$$

The **SI units of density** are kg per cubic meter.

$$1 \text{ g/cm}^3 = 1\,000 \text{ kg/m}^3$$

Pressure is defined as the normal force per unit area acting on a surface. Pressure has units of pascals (Pa).

$$P \equiv \frac{F}{A} \tag{9.2}$$

$$1 \text{ Pa} = 1 \text{ N/m}^2$$

Other pressure units include atmospheres, Torr, and pounds per sq. inch.

$$1 \text{ atm} = 1.013 \times 10^5 \text{ Pa}$$

$$1 \text{ Torr} = 133.3 \text{ Pa}$$

$$1 \text{ lb/in}^2 = 6.895 \times 10^3 \text{ Pa}$$

The elastic modulus, defined as the ratio of the *stress* to *strain*, is a general characterization of the deformation of a material. Stress is a quantity that is proportional to the force which causes the deformation, and strain is a measure of the degree of deformation. *There is an elastic modulus corresponding to each type of deformation: change in length, shape, and volume.*

$$\text{stress} = \text{elastic modulus} \times \text{strain} \tag{9.3}$$

Young's modulus (Y) is a measure of the resistance of a body to elongation or compression. It is defined as the ratio of tensile stress to tensile strain. The *elastic limit* is the maximum stress from which a substance will recover to an initial length.

$$\frac{F}{A} = Y\frac{\Delta L}{L_0} \tag{9.5}$$

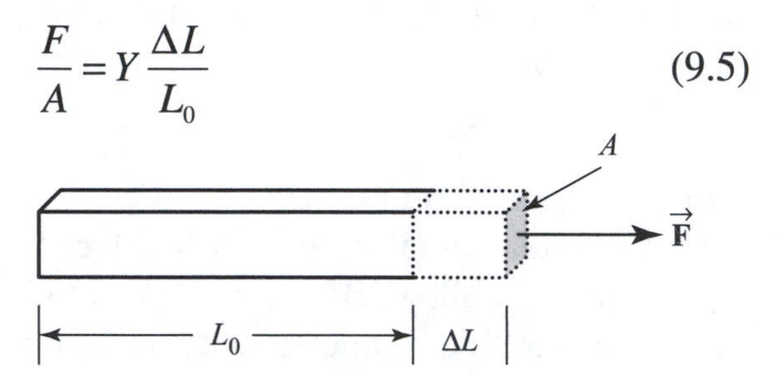

The **shear modulus** is a measure of the resistance of a material to internal planes sliding past each other. Deformation occurs when a force is applied along a direction parallel to one surface of a body. As illustrated in the figure, a force $\vec{\mathbf{F}}$ applied parallel to the top surface, with the bottom surface fixed, will cause the top surface to move forward a distance Δx.

$$\frac{F}{A} = S\frac{\Delta x}{h} \tag{9.6}$$

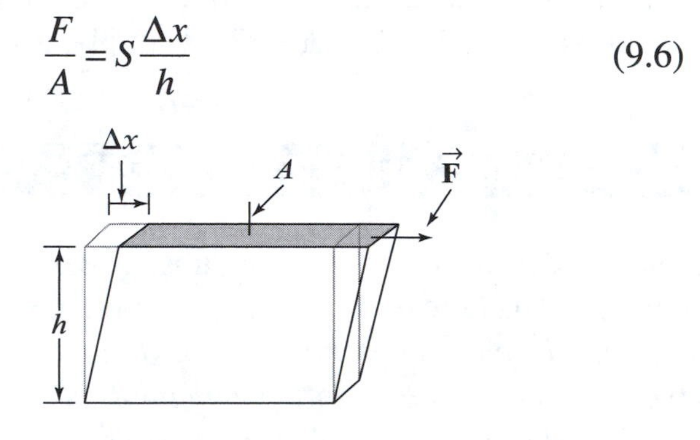

The **bulk modulus** is a measure of the resistance of a substance to uniform pressure (or squeezing) on all sides. Note that when ΔP is positive (increase in pressure), the ratio $\Delta V/V$ will be negative (decrease in volume), and vice versa. Therefore, the negative sign in the equation ensures that B will always be positive. *Compressibility of a substance is the reciprocal of the bulk modulus.*

$$\Delta P = -B\frac{\Delta V}{V} \tag{9.7}$$

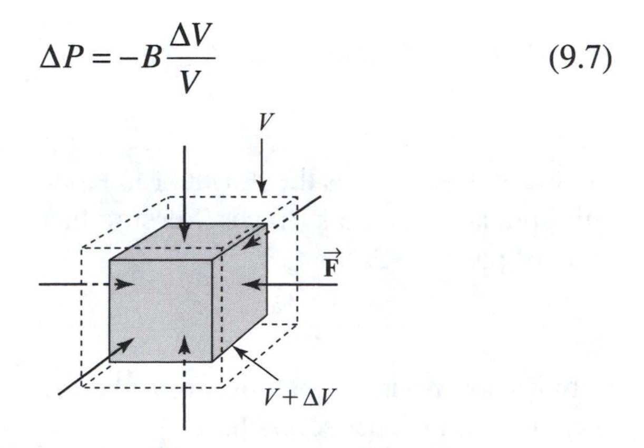

The **absolute pressure**, P, at a depth, h, below the surface of a liquid of density ρ that is open to the atmosphere is greater than atmospheric pressure, P_0, by an amount which depends on the depth below the surface. *The pressure at a given depth below the surface of a liquid has the same value at all points and does not depend on the shape of the container.*

$$P = P_0 + \rho g h \qquad (9.11)$$

$$P_0 = 1.013 \times 10^5 \text{ Pa}$$

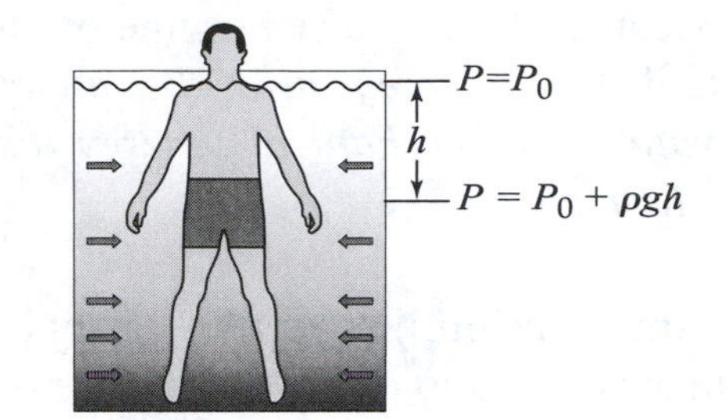

The quantity $\rho g h$ is called the *gauge pressure* and P is the *absolute pressure.*

absolute pressure = atmospheric pressure + gauge pressure

Pascal's principle states that pressure applied to an *enclosed fluid* (liquid or gas) is *transmitted* undiminished to every point within the fluid and over the walls of the vessel which contain the fluid.

Archimedes' principle states that when an object is partially or fully submerged in a fluid, the fluid exerts an upward *buoyant force, B*, which depends on the fluid density and volume of the displaced fluid. *The buoyant force equals the weight of the displaced fluid.*

$$B = \rho_{\text{fluid}} V_{\text{fluid}} g \qquad (9.12b)$$

V_{fluid} = volume of *displaced* fluid

The **ideal fluid model** is based on the following characteristics:

- **nonviscous**—internal friction between adjacent fluid layers is negligible.
- **steady flow**—the velocity, density, and pressure at each point in the fluid are constant in time.
- **incompressible**—the density throughout the fluid is constant.
- **irrotational** (without turbulence)—there are no eddy currents within the fluid (each element of the fluid has zero angular velocity about its center).

The **equation of continuity** states that *the flow rate is constant at every point along a pipe carrying an incompressible fluid.*

$$A_1 v_1 = A_2 v_2 \qquad (9.15)$$

Av = flow rate

Bernoulli's equation states that the sum of pressure, kinetic energy per unit volume, and potential energy per unit volume remains constant along a streamline of an ideal fluid. *It is a statement of the law of conservation of mechanical energy as applied to a fluid.*

$$P + \tfrac{1}{2}\rho v^2 + \rho g y = \text{constant} \qquad (9.17)$$

The **surface tension**, γ, in a film of liquid is defined as the ratio of the magnitude of the surface tension force, F, to the length, L, along which the force acts.

$$\gamma \equiv \frac{F}{L} \qquad (9.19)$$

In a **capillary tube**, liquid will rise when adhesive forces are greater than cohesive forces (e.g. water and glass as shown in the figure on the left); they will be depressed relative to the surface when the cohesive forces are greater (e.g. mercury and glass as shown in the figure on the right).

$$h = \frac{2\gamma}{\rho g r}\cos\phi \qquad (9.22)$$

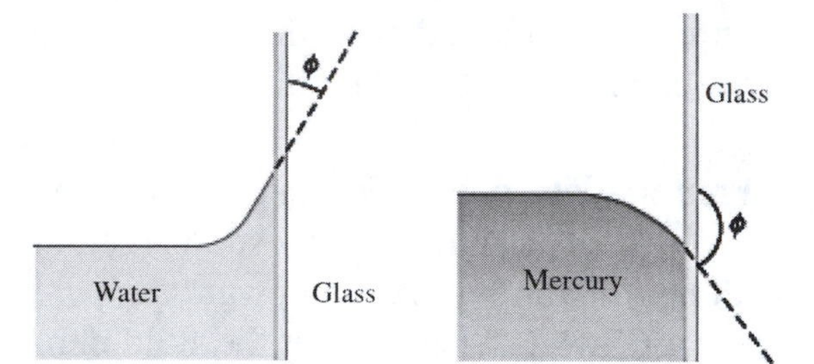

The coefficient of viscosity, η (the lowercase Greek letter eta), may be thought of as the ratio of the shearing stress to the *rate of change* of the shear strain. The SI unit of viscosity is the poise. Consider a layer of liquid of thickness d and area A between two solid surfaces. In Equation 9.23, F is the force required to move one of the solid surfaces with a velocity v relative to the other.

$$F = \eta \frac{Av}{d} \qquad (9.23)$$

$$1 \text{ poise} = 10^{-1}\ \text{N}\cdot\text{s/m}^2 \qquad (9.24)$$

The **Reynolds number** is a dimensionless factor which is an indicator of the flow characteristics of a liquid in a tube. It is a function of the density of the fluid, the average speed of the fluid along the direction of flow, the diameter of the tube, and the viscosity of the fluid.

$$RN = \frac{\rho v d}{\eta} \qquad (9.26)$$

Streamline Flow: $RN < \sim 2\,000$

Unstable Flow: $2\,000 < RN < 3\,000$

Turbulent Flow: $RN > 3\,000$

The **diffusion rate** is a measure of the mass being transported per unit time. Equation 9.27 is called Fick's law.

$$\text{Diffusion Rate} = \frac{\text{mass}}{\text{time}} = \frac{\Delta M}{\Delta t} = DA\left(\frac{C_2 - C_1}{L}\right) \quad (9.27)$$

D = diffusion coefficient

$(C_2 - C_1)/L$ = concentration gradient

Stokes' law describes the resistive force on a small spherical object of radius r falling with speed v through a viscous fluid.

$$F_r = 6\pi\eta r v \quad (9.28)$$

η = coefficient of viscosity

Terminal speed is achieved as a sphere of radius r falls through a viscous medium. The sphere is acted on by three forces: the force of frictional resistance, the buoyant force of the fluid, and the weight of the sphere. When the net upward force balances the downward weight force, the sphere reaches terminal speed.

$$v_t = \frac{2r^2 g}{9\eta}(\rho - \rho_f) \quad (9.29)$$

v_t = terminal speed.

In a **centrifuge** those particles having the greatest mass will have the largest terminal speed. Therefore, the most massive particles will settle out of the mixture. The factor k is a coefficient of frictional resistance which must be determined experimentally.

$$v_t = \frac{m\omega^2 r}{k}\left(1 - \frac{\rho_f}{\rho}\right) \quad (9.32)$$

ω = angular speed

$\omega^2 r$ = radial acceleration

REVIEW CHECKLIST

- Describe the three types of deformations that can occur in a solid, and define the elastic modulus that is used to characterize each (Young's modulus, shear modulus, and bulk modulus).
- Understand the concept of pressure at a point in a fluid, and the variation of pressure with depth. Understand the relationships among absolute, gauge, and atmospheric pressure values; and know the several different units commonly used to express pressure.
- Understand the origin of buoyant forces; and state and explain Archimedes' principle.
- State and understand the physical significance of the equation of continuity (constant flow rate) and Bernoulli's equation for fluid flow (relating flow velocity, pressure, and pipe elevation).

SOLUTIONS TO SELECTED END-OF-CHAPTER PROBLEMS

7. Suppose a distant world with surface gravity of 7.44 m/s^2 has an atmospheric pressure of 8.04×10^4 Pa at the surface. (a) What force is exerted by the atmosphere on a disk-shaped region 2.00 m in radius at the surface of a methane ocean? (b) What is the weight of a 10.0-m-deep cylindrical column of methane with radius 2.00 m? (c) Calculate the pressure at a depth of 10.0 m in the methane ocean. *Note:* The density of liquid methane is 415 kg/m^3.

Solution

(a) With atmospheric pressure at the ocean surface of $P_{\text{atm}} = 8.04 \times 10^4$ Pa, the downward force the atmosphere exerts on a circular area 2.00 m in radius on the ocean surface is

$$F_{\text{atm}} = P_{\text{atm}} A = \left(8.04 \times 10^4 \text{ Pa}\right)\left[\pi(2.00 \text{ m})^2\right] = 1.01 \times 10^6 \text{ N} \quad \Diamond$$

(b) The weight of a cylindrical column of methane having a radius of $r = 2.00$ m and height $h = 10.0$ m is $F_g = mg = (\rho V)g$. With surface gravity of $g = 7.44 \text{ m/s}^2$ on this planet and a density of $\rho = 415 \text{ kg/m}^3$ for liquid methane, this yields a weight of

$$F_g = \left(415 \text{ kg/m}^3\right)\left[\pi(2.00 \text{ m})^2(10.0 \text{ m})\right]\left(7.44 \text{ m/s}^2\right) = 3.88 \times 10^5 \text{ N} \quad \Diamond$$

(c) To compute the pressure at a depth of 10.0 m in the methane ocean, we consider the circular area of 2.00-m radius forming the bottom of the cylindrical column of methane considered in part (b) above. The total downward force acting on this area equals the weight F_g of the methane column plus the downward force F_{atm} exerted on the top of the methane column by the atmosphere above. Thus, the force per unit area or pressure on our circular area is

$$P = \frac{F_{\text{total}}}{A} = \frac{F_g + F_{\text{atm}}}{\pi r^2} = \frac{3.88 \times 10^5 \text{ N} + 1.01 \times 10^6 \text{ N}}{\pi(2.00 \text{ m})^2} = 1.11 \times 10^5 \text{ Pa} \quad \Diamond$$

As shown in Section 9.4, the pressure at this depth in the methane ocean may also be computed from $P = P_0 + \rho g h$, where P_0 is the pressure at some reference level in a fluid and h is the depth we are below that reference level. Choosing our reference level at the ocean surface where $P_0 = P_{\text{atm}}$, the pressure at a depth of 10.0 m in the ocean is then

$$P = P_{\text{atm}} + \rho g h = 8.04 \times 10^4 \text{ Pa} + \left(415 \text{ kg/m}^3\right)\left(7.44 \text{ m/s}^2\right)(10.0 \text{ m}) = 1.11 \times 10^5 \text{ Pa} \quad \Diamond$$

13. For safety in climbing, a mountaineer uses a nylon rope that is 50 m long and 1.0 cm in diameter. When supporting a 90-kg climber, the rope elongates 1.6 m. Find its Young's modulus.

Solution

Young's modulus of a material is a measure of how effectively that material can resist attempts to distort its length by either tension or compression forces. It is the ratio of the tensile stress to the tensile strain. The tensile stress is defined as $stress = F/A$, where F is the magnitude of the distorting force and A is the cross-sectional area of the material; while the tensile strain is given by $strain = \Delta L/L_0$, where ΔL is the change in length produced by the distorting force and L_0 is the relaxed length of the material. Thus,

$$Y = \frac{F/A}{\Delta L/L_0} = \frac{F \cdot L_0}{\Delta L \cdot A}$$

For the mountaineer's climbing rope, the distorting force is the mountaineer's weight $F = mg$, and the cross-sectional area is $A = \pi d^2/4$, where d is the diameter of the rope. With an initial length of $L_0 = 50$ m, and an elongation of $\Delta L = 1.6$ m, Young's modulus of the nylon rope material is found to be

$$Y = \frac{mg \cdot L_0}{\Delta L \cdot \pi d^2/4} = \frac{4(90 \text{ kg})(9.80 \text{ m/s}^2)(50 \text{ m})}{(1.6 \text{ m})\pi(1.0 \times 10^{-2} \text{ m})^2} = 3.5 \times 10^8 \text{ N/m}^2 = 3.5 \times 10^8 \text{ Pa} \quad \Diamond$$

---

19. Determine the elongation of the rod in Figure P9.19 if it is under a tension of 5.8×10^3 N.

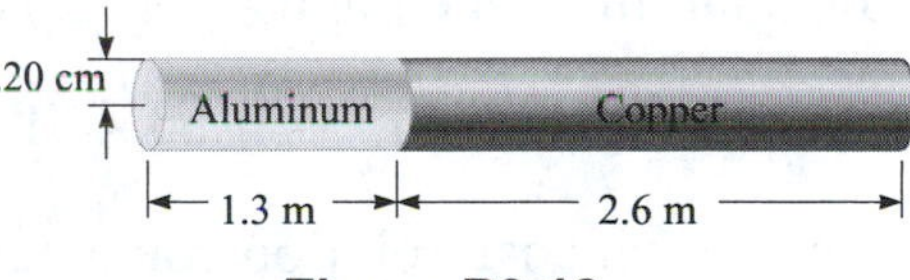

Figure P9.19

Solution

When the rod comes to equilibrium, the tension will be uniform throughout its length, with both types of metal subject to a stretching force of $F = 5.8 \times 10^3$ N. Also, the cross-sectional area is uniform throughout the rod, with a value of

$$A = \pi r^2 = \pi(0.20 \times 10^{-2} \text{ m})^2 = \pi(2.0 \times 10^{-3} \text{ m})^2 = 4.0\pi \times 10^{-6} \text{ m}^2$$

The total elongation of the rod is the sum of the elongation of the aluminum section and that of the copper section, or $\Delta L_{rod} = \Delta L_{Al} + \Delta L_{Cu}$.

Young's modulus of a material is given by

$$Y = \frac{tensile\ stress}{tensile\ strain} = \frac{F/A}{\Delta L/L_0}$$

Here, F is the force of tension or compression acting on the material, A is the cross-sectional area, L_0 is the original length of the material, and ΔL is the elongation the material undergoes. Hence, the elongation of each section of this rod is given by $\Delta L = FL_0/YA$, and the total elongation of the rod becomes

$$\Delta L_{\text{rod}} = \frac{F(L_0)_{\text{Al}}}{Y_{\text{Al}}A} + \frac{F(L_0)_{\text{Cu}}}{Y_{\text{Cu}}A} = \frac{F}{A}\left[\frac{(L_0)_{\text{Al}}}{Y_{\text{Al}}} + \frac{(L_0)_{\text{Cu}}}{Y_{\text{Cu}}}\right]$$

or

$$\Delta L_{\text{rod}} = \frac{5.8\times10^3\ \text{N}}{4.0\pi\times10^{-6}\ \text{m}^2}\left[\frac{1.3\ \text{m}}{7.0\times10^{10}\ \text{Pa}} + \frac{2.6\ \text{m}}{11\times10^{10}\ \text{Pa}}\right] = 1.9\times10^{-2}\ \text{m} = 1.9\ \text{cm} \quad \lozenge$$

---

23. A collapsible plastic bag (Figure P9.23) contains a glucose solution. If the average gauge pressure in the vein is 1.33×10^3 Pa, what must be the minimum height h of the bag in order to infuse glucose into the vein? Assume the specific gravity of the solution is 1.02.

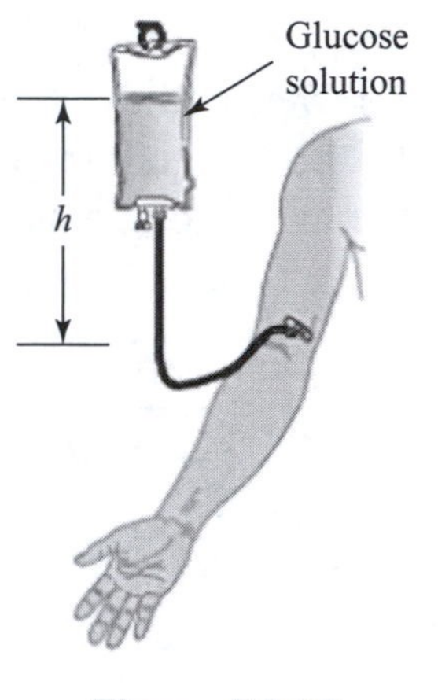

Figure P9.23

Solution

If the glucose solution has a specific gravity of 1.02, its density is

$$\rho_{\text{gs}} = 1.02\cdot\rho_{\text{water}} = 1.02\left(1.00\times10^3\ \text{kg/m}^3\right) = 1.02\times10^3\ \text{kg/m}^3$$

The gauge pressure in the vein is $\left(P_{\text{gauge}}\right)_{\text{vein}} = 1.33\times10^3$ Pa, so in order to infuse glucose into the vein, the gauge pressure in the glucose solution at the point where it is to enter the vein must be

$$\left(P_{\text{gauge}}\right)_{\text{gs}} \geq \left(P_{\text{gauge}}\right)_{\text{vein}} = 1.33\times10^3\ \text{Pa}$$

Since the glucose solution container is a collapsible plastic bag, the bag will collapse and maintain atmospheric pressure at the upper surface of the solution as the solution drains out of the bag. The pressure at depth h below this upper surface (i.e., at the level where the glucose enters the vein) is given by $P = P_{\text{atm}} + \rho_{\text{gs}}gh$, and the gauge pressure at the entry point to the vein is $\left(P_{\text{gauge}}\right)_{\text{gs}} = P - P_{\text{atm}} = \rho_{\text{gs}}gh$. Thus, for infusion into the vein, it is necessary to have $\rho_{\text{gs}}gh \geq 1.33\times10^3$ Pa or

$$h \geq \frac{1.33\times10^3\ \text{Pa}}{\rho_{\text{gs}}g} = \frac{1.33\times10^3\ \text{Pa}}{\left(1.02\times10^3\ \text{kg/m}^3\right)\left(9.80\ \text{m/s}^2\right)} = 0.133\ \text{m} \quad \lozenge$$

---

27. Figure P9.27 shows the essential parts of a hydraulic brake system. The area of the piston in the master cylinder is 1.8 cm^2 and that of the piston in the brake cylinder is 6.4 cm^2. The coefficient of friction between shoe and wheel drum is 0.50. If the wheel has a radius of 34 cm, determine the frictional torque about the axle when a force of 44 N is exerted on the brake pedal.

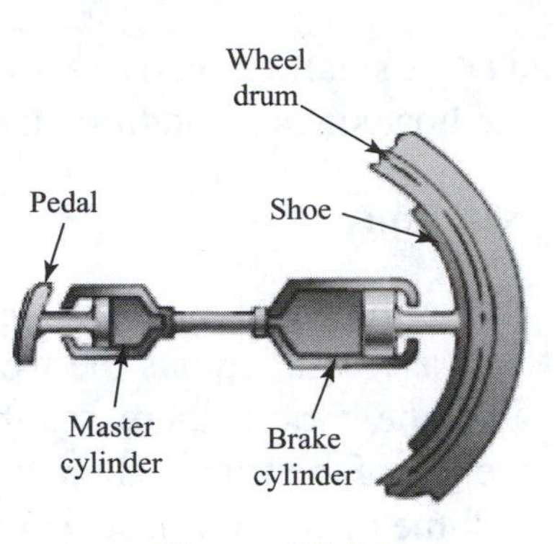

Figure P9.27

Solution

The brake pedal exerts a force of $F_1 = 44$ N on the piston having area A_1 in the master cylinder. This creates an increase in pressure of

$$\Delta P = \frac{F_{\text{pedal}}}{A_{\substack{\text{Master}\\\text{cylinder}}}}$$

in the enclosed fluid of the hydraulic brake system. According to Pascal's principle, this change in pressure is transmitted undiminished to every point of the fluid and to the walls of the container. Thus, an increase in pressure also occurs in the fluid inside the brake cylinder and this fluid will exert an increased force of

$$F_{\text{Shoe}} = (\Delta P)A_{\substack{\text{Brake}\\\text{cylinder}}} = \left(\frac{F_{\text{pedal}}}{A_{\substack{\text{Master}\\\text{cylinder}}}}\right)A_{\substack{\text{Brake}\\\text{cylinder}}} = \left(\frac{44\text{ N}}{1.8\ \cancel{\text{cm}^2}}\right)\left(6.4\ \cancel{\text{cm}^2}\right) = 1.6\times10^2\text{ N}$$

on the piston of the brake cylinder, and hence as a normal force to the brake shoe. The friction force between the brake shoe and the wheel drum will then be

$$f_k = \mu_k n = \mu_k F_{\text{Shoe}} = (0.50)\left(1.6\times10^2\text{ N}\right) = 80\text{ N}$$

This friction force between the shoe and drum will produce a frictional torque about the axle of magnitude

$$\tau = f_k \cdot r = (80\text{ N})(0.34\text{ m}) = 27\text{ N}\cdot\text{m}$$ ◊

---

31. A small ferryboat is 4.00 m wide and 6.00 m long. When a loaded truck pulls onto it, the boat sinks an additional 4.00 cm into the river. What is the weight of the truck?

Solution

According to Archimedes' principle, the floating ferryboat must displace a volume of water whose weight equals the weight of the boat and its contents. When the loaded truck pulls onto the boat, it increases the weight of the boat and its load by an amount equal to the weight of the truck. Thus, the boat must sink lower into the water to displace an additional volume of water whose weight equals the weight of the truck.

If the cross-sectional area of the bottom of the boat is

$$A = (4.00\text{ m})(6.00\text{ m}) = 24.0\text{ m}^2$$

and the boat sinks an additional distance $h = 4.00$ cm into the water when the truck pulls onto it, the additional volume of water displaced is

$$\Delta V = A \cdot h = \left(24.0\text{ m}^2\right)\left(4.00 \times 10^{-2}\text{ m}\right) = 0.960\text{ m}^3$$

Since water has a density of $\rho_{\text{water}} = 1.00 \times 10^3\ \text{kg/m}^3$, the weight of this additional displaced water, and hence the weight of the loaded truck, is

$$W = (\Delta m)g = \left(\rho_{\text{water}}\Delta V\right)g = \left[\left(1.00 \times 10^3\ \text{kg/m}^3\right)\left(0.960\text{ m}^3\right)\right]\left(9.80\text{ m/s}^2\right)$$

or

$$W = 9.41 \times 10^3\text{ N} = 9.41\text{ kN}$$ ◊

40. A light spring of force constant $k = 160$ N/m rests vertically on the bottom of a large beaker of water (Fig. P9.40a). A 5.00-kg block of wood (density = 650 kg/m^3) is connected to the spring, and the block–spring system is allowed to come to static equilibrium (Fig. P9.40b). What is the elongation ΔL of the spring?

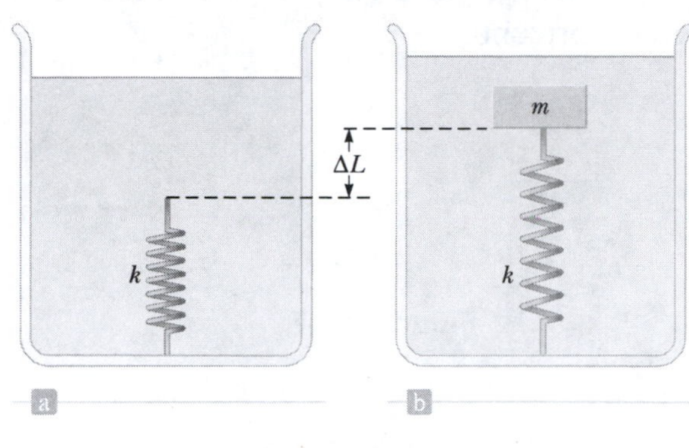

Figure P9.40

Solution

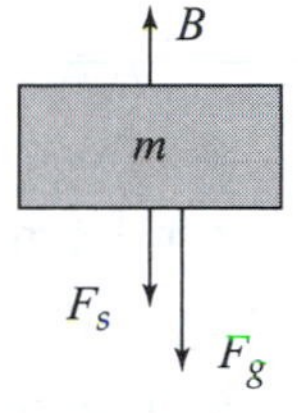

When the block is held in equilibrium on the end of the spring, three forces act on it as shown in the sketch at the right. These are a downward gravitational force, $F_g = mg$, a downward spring force of magnitude $F_s = k\Delta L$, and an upward buoyancy force whose magnitude equals the weight of the water displaced by the block.

If the block is made from wood having density $\rho_{\text{wood}} = 650$ kg/m^3, its volume must be given by $V = m/\rho_{\text{wood}}$, where its mass is $m = 5.00$ kg. The weight of the displaced water, and the magnitude of the buoyancy force, is then

$$B = (\rho_{\text{water}} V)g = (\rho_{\text{water}}/\rho_{\text{wood}})mg$$

Since the block is held in equilibrium by the forces acting on it, Newton's second law requires that

$$\Sigma F_y = B - F_s - F_g = 0 \qquad \text{or} \qquad F_s = B - F_g$$

which gives $k\Delta L = (\rho_{\text{water}}/\rho_{\text{wood}})mg - mg = [(\rho_{\text{water}}/\rho_{\text{wood}}) - 1]mg$. This gives the extension of the spring at equilibrium as

$$\Delta L = \frac{1}{k}\left[\left(\frac{\rho_{\text{water}}}{\rho_{\text{wood}}}\right) - 1\right]mg$$

or

$$\Delta L = \frac{1}{(160\ \text{N/m})}\left[\left(\frac{1.00\times 10^3\ \text{kg/m}^3}{650\ \text{kg/m}^3}\right) - 1\right](5.00\ \text{kg})(9.80\ \text{m/s}^2) = 0.165\ \text{m} = 16.5\ \text{cm} \quad \lozenge$$

47. A hypodermic syringe contains a medicine with the density of water (Fig. P9.47). The barrel of the syringe has a cross-sectional area of 2.50×10^{-5} m^2. In the absence of a force on the plunger, the pressure everywhere is 1.00 atm. A force $\vec{\mathbf{F}}$ of magnitude 2.00 N is exerted on the plunger, making medicine squirt from the needle. Determine the medicine's flow speed through the needle. Assume the pressure in the needle remains equal to 1.00 atm and that the syringe is horizontal.

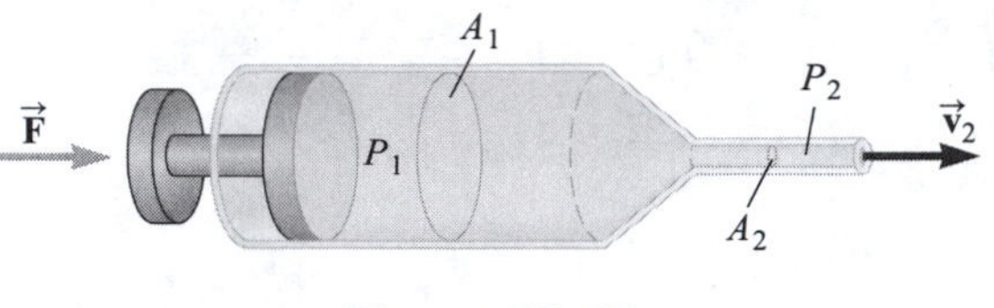

Figure P9.47

Solution

The pressure inside the needle, which is open to the atmosphere, remains at $P_2 = 1.00$ atm as force is applied to the plunger of the needle. Applying a force of $F = 2.00$ N to the plunger produces an increase in pressure inside the syringe chamber of

$$\Delta P = P_1 - 1.00 \text{ atm} = P_1 - P_2 = \frac{F}{A_1} = \frac{2.00 \text{ N}}{2.50 \times 10^{-5} \text{ m}^2} = 8.00 \times 10^4 \text{ Pa}$$

From Bernoulli's equation, $P_1 + \frac{1}{2}\rho_{\text{water}} v_1^2 + \rho_{\text{water}} g y_1 = P_2 + \frac{1}{2}\rho_{\text{water}} v_2^2 + \rho_{\text{water}} g y_2$, with $y_1 = y_2$ since the syringe is horizontal, and $v_1 \approx 0$ since the diameter of the syringe chamber is much larger than that of the opening to the needle. This gives

$$v_2^2 = v_1^2 + \frac{2(P_1 - P_2)}{\rho_{\text{water}}} + 2g(y_1 - y_2) \approx 0 + \frac{2(P_1 - P_2)}{\rho_{\text{water}}} + 0$$

or the medicine's flow speed through the needle is

$$v_2 = \sqrt{\frac{2(P_1 - P_2)}{\rho_{\text{water}}}} = \sqrt{\frac{2(8.00 \times 10^4 \text{ Pa})}{1.00 \times 10^3 \text{ kg/m}^3}} = 12.6 \text{ m/s} \qquad \Diamond$$

58. The Venturi tube shown in Figure P9.58 may be used as a fluid flowmeter. Suppose the device is used at a service station to measure the flow rate of gasoline ($\rho = 7.00 \times 10^2\ \text{kg/m}^3$) through a hose having an outlet radius of 1.20 cm. If the difference in pressure is measured to be $P_1 - P_2 = 1.20$ kPa and the radius of the inlet tube to the meter is 2.40 cm, find (a) the speed of the gasoline as it leaves the hose and (b) the fluid flow rate in cubic meters per second.

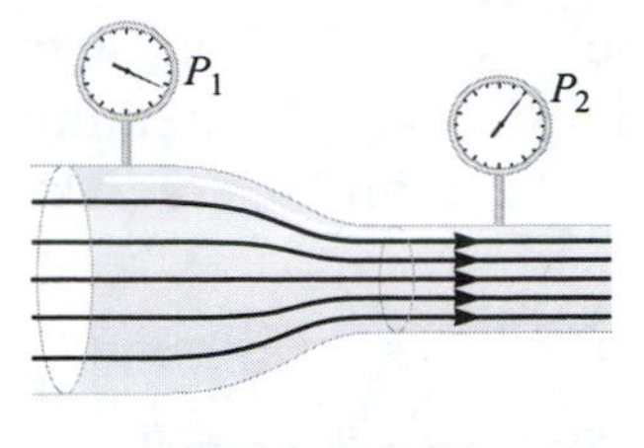

Figure P9.58

Solution

(a) The pressure, flow speed, and elevation at different points in a flowing fluid having density ρ are related by Bernoulli's equation which states

$$P_1 + \tfrac{1}{2}\rho v_1^2 + \rho g y_1 = P_2 + \tfrac{1}{2}\rho v_2^2 + \rho g y_2$$

Since this tube is horizontal, $y_2 = y_1$, and the gravity terms cancel from the equation, leaving $P_1 + \tfrac{1}{2}\rho v_1^2 = P_2 + \tfrac{1}{2}\rho v_2^2$. This may be rearranged to give

$$v_2^2 - v_1^2 = \frac{2(P_1 - P_2)}{\rho} = \frac{2(1.20 \times 10^3\ \text{Pa})}{7.00 \times 10^2\ \text{kg/m}^3}$$

or

$$v_2^2 - v_1^2 = 3.43\ \text{m}^2/\text{s}^2 \qquad \textbf{[1]}$$

From the continuity equation, $A_1 v_1 = A_2 v_2$, we obtain

$$v_2 = \left(\frac{A_1}{A_2}\right) v_1 = \left(\frac{\pi r_1^2}{\pi r_2^2}\right) v_1 = \left(\frac{r_1}{r_2}\right)^2 v_1 = \left(\frac{2.40\ \text{cm}}{1.20\ \text{cm}}\right)^2 v_1$$

or $v_2 = 4v_1$. Substituting this result into Equation [1] gives $15v_1^2 = 3.43\ \text{m}^2/\text{s}^2$ and $v_1 = 0.478\ \text{m/s}$. Then, the flow speed of the gasoline as it leaves the hose is

$$v_2 = 4v_1 = 4(0.478\ \text{m/s}) = 1.91\ \text{m/s} \qquad \lozenge$$

(b) The volume flow rate of gasoline through the hose is

$$\text{Volume flow rate} = A_2 v_2 = \left(\pi r_2^2\right) v_2$$

$$= \pi\left(1.20 \times 10^{-2}\ \text{m}\right)^2 (1.91\ \text{m/s}) = 8.64 \times 10^{-4}\ \text{m}^3/\text{s} \qquad \lozenge$$

67. Spherical particles of a protein of density 1.8 g/cm³ are shaken up in a solution of 20°C water. The solution is allowed to stand for 1.0 h. If the depth of water in the tube is 5.0 cm, find the radius of the largest particles that remain in solution at the end of the hour.

Solution

Note that the density of the protein is

$$\rho = 1.8\ \frac{\text{g}}{\text{cm}^3}\left(\frac{1\ \text{kg}}{10^3\ \text{g}}\right)\left(\frac{10^2\ \text{cm}}{1\ \text{m}}\right)^3 = 1.8 \times 10^3\ \text{kg/m}^3$$

If a particle is still in suspension after one hour, this means that it has fallen less than 5.0 cm in the elapsed time. Then its terminal speed must be

$$v_t < 5.0\ \text{cm/h} = \left(\frac{5.0\ \text{cm}}{1.0\ \text{h}}\right)\left(\frac{1\ \text{m}}{100\ \text{cm}}\right)\left(\frac{1\ \text{h}}{3\,600\ \text{s}}\right) = 1.4 \times 10^{-5}\ \text{m/s}$$

In general, the terminal speed of a particle of density ρ and radius r falling through a fluid of density ρ_f and viscosity η is given by Equation 9.29 in the textbook as

$$v_t = \frac{2r^2 g}{9\eta}\left(\rho - \rho_f\right)$$

Thus, if the upper limit of the terminal velocity of the particles still suspended in the 20°C water is $v_{t,\ \text{max}} = 1.4 \times 10^{-5}$ m/s, the maximum radius of these particles is

$$r_{\text{max}} = \sqrt{\frac{9\eta\, v_{t,\ \text{max}}}{2g\left(\rho - \rho_f\right)}} = \sqrt{\frac{9\left(1.0 \times 10^{-3}\ \text{N}\cdot\text{s/m}^2\right)\left(1.4 \times 10^{-5}\ \text{m/s}\right)}{2\left(9.8\ \text{m/s}^2\right)\left[(1.8 - 1.0) \times 10^3\ \text{kg/m}^3\right]}}$$

or

$$r_{\text{max}} = 2.8 \times 10^{-6}\ \text{m} = 2.8\ \mu\text{m}$$ ◊

70. Water is forced out of a fire extinguisher by air pressure, as shown in Figure P9.70. What gauge air pressure in the tank (above atmospheric pressure) is required for the water to have a jet speed of 30.0 m/s when the water level in the tank is 0.500 m below the nozzle?

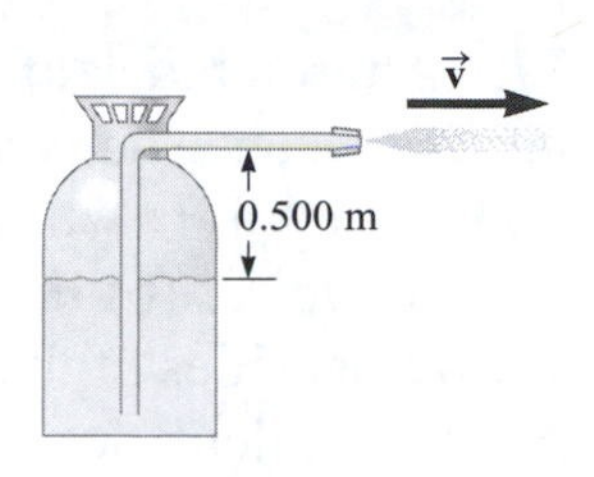

Figure P9.70

Solution

Bernoulli's equation, which states

$$P_1 + \tfrac{1}{2}\rho v_1^2 + \rho g y_1 = P_2 + \tfrac{1}{2}\rho v_2^2 + \rho g y_2$$

gives the relation among the pressure, kinetic energy per unit volume, and potential energy per unit volume at different points in a flowing fluid of density ρ. For the fire extinguisher shown in Figure P9.70, we take point 1 at the surface of the water in the tank and point 2 at the point where the water emerges from the nozzle. Then, since the nozzle is open to the atmosphere, $P_2 = P_{atm}$, and Bernoulli's equation gives the gauge pressure of the air in the tank as

$$P_{gauge} = P_1 - P_{atm} = P_1 - P_2 = \tfrac{1}{2}\rho\left(v_2^2 - v_1^2\right) + \rho g\left(y_2 - y_1\right)$$

Because the volume of water in the tank is much larger than what emerges from the nozzle each second, v_1^2 will be very small in comparison to v_2^2. We then approximate $v_2^2 - v_1^2 \approx v_2^2$, and find the required gauge pressure as

$$P_{gauge} = \rho\left[\frac{v_2^2}{2} + g\left(y_2 - y_1\right)\right]$$

With water having a density of $\rho = 1.00 \times 10^3\ \text{kg/m}^3$, the gauge pressure needed to give a jet speed of $v_2 = 30.0\ \text{m/s}$ at the nozzle when $y_2 - y_1 = 0.500\ \text{m}$ is

$$P_{gauge} = \left(1.00 \times 10^3\ \text{kg/m}^3\right)\left[\frac{(30.0\ \text{m/s})^2}{2} + \left(9.80\ \text{m/s}^2\right)(0.500\ \text{m})\right]$$

or

$$P_{gauge} = 4.55 \times 10^5\ \text{Pa} = 455\ \text{kPa}$$ ◊

77. An iron block of volume 0.20 m^3 is suspended from a spring scale and immersed in a flask of water. Then the iron block is removed, and an aluminum block of the same volume replaces it. (a) In which case is the buoyant force the greatest, for the iron block or the aluminum block? (b) In which case does the spring scale read the largest value? (c) Use the known densities of these materials to calculate the quantities requested in parts (a) and (b). Are your calculations consistent with your previous answers to parts (a) and (b)?

Solution

(a) The buoyant force acting on an object partially or wholly submerged in a fluid of density ρ_f is equal to the weight of the displaced fluid, or $B = \rho_f g V_{\text{displaced}}$. Both iron and aluminum are denser than water, so both blocks will be fully submerged. Since the two blocks have the same volume, they displace equal volumes of water, and will experience buoyant forces of equal magnitudes. ◊

(b) The force exerted on the block by the spring scale (of magnitude equal to the scale reading) and the buoyant force act together to support the weight of the block $\left(\text{i.e., } F_s + B = w_{\text{block}} \text{ giving } F_s = w_{\text{block}} - B\right)$. Since the volumes and the buoyant forces are the same for the two blocks, the higher spring scale reading occurs for the block having the higher density, and hence greater weight. This is the iron block, with $\rho_{\text{Fe}} > \rho_{\text{Al}}$. ◊

(c) The buoyant force in each case is

$$B = \rho_{\text{water}} g V_{\text{block}} = \left(1.00 \times 10^3 \text{ kg/m}^3\right)\left(9.80 \text{ m/s}^2\right)\left(0.20 \text{ m}^3\right) = 2.0 \times 10^3 \text{ N} \quad ◊$$

The scale reading for the iron block is $\left.F_s\right)_{\text{Fe}} = m_{\text{Fe}} g - B = \rho_{\text{Fe}} g V - B$, or

$$\left.F_s\right)_{\text{Fe}} = \left(7.86 \times 10^3 \text{ kg/m}^3\right)\left(9.8 \text{ m/s}^2\right)\left(0.20 \text{ m}^3\right) - 2.0 \times 10^3 \text{ N} = 1.3 \times 10^4 \text{ N} \quad ◊$$

while that for the aluminum block is $\left.F_s\right)_{\text{Al}} = m_{\text{Al}} g - B = \rho_{\text{Al}} g V - B$, or

$$\left.F_s\right)_{\text{Al}} = \left(2.70 \times 10^3 \text{ kg/m}^3\right)\left(9.8 \text{ m/s}^2\right)\left(0.20 \text{ m}^3\right) - 2.0 \times 10^3 \text{ N} = 3.3 \times 10^3 \text{ N} \quad ◊$$

all of which is consistent with previous answers in parts (a) and (b).

86. A helium-filled balloon, whose envelope has a mass of 0.25 kg, is tied to a 2.0-m-long, 0.050-kg string. The balloon is spherical with a radius of 0.40 m. When released, it lifts a length h of the string and then remains in equilibrium, as in Figure P9.86. Determine the value of h. [*Hint:* Only that part of the string above the floor contributes to the load being supported by the balloon.]

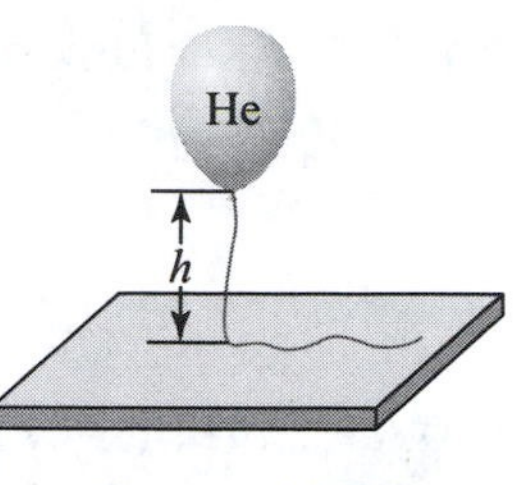

Figure P9.86

Solution

The upward buoyant force acting on the balloon is equal to the weight of air displaced by the balloon, or $B = \rho_{\text{air}} g V_{\text{balloon}}$. In addition to this force, there are three downward forces acting on the balloon. These are: (1) the weight of the elastic material making up the balloon itself, $w_{\text{balloon}} = m_{\text{balloon}} g$; (2) the weight of the helium filling the balloon, $w_{\text{He}} = m_{\text{He}} g = \rho_{\text{He}} g V_{\text{balloon}}$; and (3) the weight of the length h of string the balloon has lifted off the floor, $w_{\text{string}} = m_{\substack{\text{lifted}\\\text{string}}} g = \left(\lambda_{\text{string}} h\right) g$, where λ_{string} is the mass per unit length of the string with a value of

$$\lambda_{\text{string}} = \frac{m_{\substack{\text{total}\\\text{length}}}}{L_{\text{string}}} = \frac{0.050 \text{ kg}}{2.0 \text{ m}} = 2.5 \times 10^{-2} \text{ kg/m}$$

When the balloon comes to equilibrium, $\Sigma F_y = B - w_{\text{balloon}} - w_{\text{He}} - w_{\text{string}} = 0$, or

$$\rho_{\text{air}} \cancel{g} V_{\text{balloon}} = m_{\text{balloon}} \cancel{g} + \rho_{\text{He}} \cancel{g} V_{\text{balloon}} + \left(\lambda_{\text{string}} \cancel{g}\right) h$$

and the length of string lifted above the floor will be

$$h = \frac{\left(\rho_{\text{air}} - \rho_{\text{He}}\right) V_{\text{balloon}} - m_{\text{balloon}}}{\lambda_{\text{string}}} = \frac{\left(\rho_{\text{air}} - \rho_{\text{He}}\right)\left(4\pi r_{\text{balloon}}^3/3\right) - m_{\text{balloon}}}{\lambda_{\text{string}}}$$

Using density values from Table 9.1 in the textbook yields

$$h = \frac{\left[(1.29 - 0.179) \text{ kg/m}^3\right](4\pi/3)(0.40 \text{ m})^3 - 0.25 \text{ kg}}{2.5 \times 10^{-2} \text{ kg/m}} = 1.9 \text{ m} \qquad \lozenge$$

---

91. A water tank open to the atmosphere at the top has two small holes punched in its side, one above the other. The holes are 5.00 cm and 12.0 cm above the floor. How high does water stand in the tank if the two streams of water hit the floor at the same place?

Solution

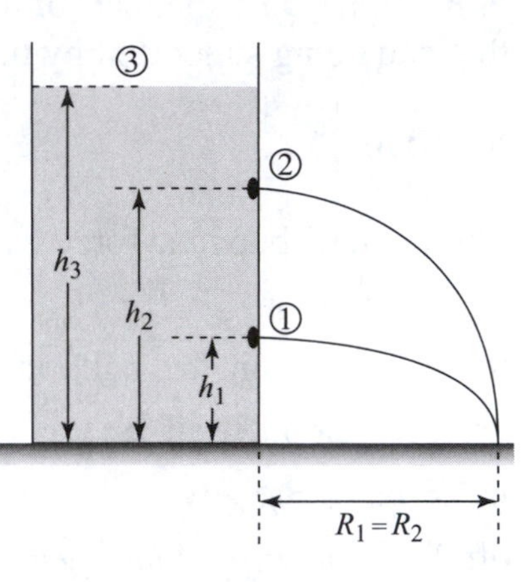

A water droplet emerging from the tank becomes a projectile with initial velocity components of $v_{0y} = 0$ and $v_{0x} = v$, where v is the speed of water passing through one of the holes in the side of the tank. The time required for this droplet to fall distance h to the floor is given by $\Delta y = v_{0y}t + a_y t^2/2$, with $\Delta y = -h$, $v_{0y} = 0$, and $a_y = -g$, is $t = \sqrt{2h/g}$. The horizontal range of the droplet is then $R = v_{0x}t = v\sqrt{2h/g}$. If the two streams are to hit the floor at the same spot, it is necessary that $R_1 = R_2$, or

$$v_1\sqrt{\frac{2h_1}{g}} = v_2\sqrt{\frac{2h_2}{g}}$$

so

$$v_1 = v_2\sqrt{\frac{h_2}{h_1}} = v_2\sqrt{\frac{12.0 \text{ cm}}{5.00 \text{ cm}}}$$

and

$$v_1^2 = 2.40 \cdot v_2^2 \qquad \textbf{[1]}$$

Recognize that the tank is open to the atmosphere at both points 1 (the lower hole) and 3 (the upper surface of the water in the tank), so that $P_1 = P_3 = P_{\text{atm}}$. Also, assuming the tank is very large in comparison to the size of the openings in its side, $v_3 \approx 0$. Then, writing Bernoulli's equation for this pair of points gives

$$\cancel{P_1} + \frac{1}{2}\rho_{\text{water}}v_1^2 + \rho_{\text{water}}gh_1 = \cancel{P_3} + \frac{1}{2}\rho_{\text{water}}(0) + \rho_{\text{water}}gh_3$$

or

$$v_1^2 = 2g(h_3 - h_1) \qquad \textbf{[2]}$$

Next, apply Bernoulli's equation to points 2 (the upper hole) and 3 (water surface), realizing that $P_2 = P_3 = P_{\text{atm}}$ and $v_3 \approx 0$, to obtain

$$\cancel{P_2} + \frac{1}{2}\rho_{\text{water}}v_2^2 + \rho_{\text{water}}gh_2 = \cancel{P_3} + \frac{1}{2}\rho_{\text{water}}(0) + \rho_{\text{water}}gh_3$$

or

$$v_2^2 = 2g(h_3 - h_2) \qquad \textbf{[3]}$$

Substituting equations [2] and [3] into [1] gives $h_3 - h_1 = 2.40(h_3 - h_2)$, and solving for h_3 (the depth of the water inside the tank) yields

$$h_3 = \frac{2.40h_2 - h_1}{1.40} = \frac{2.40(12.0 \text{ cm}) - 5.00 \text{ cm}}{1.40} = 17.0 \text{ cm}$$

◊

10

Thermal Physics

NOTES ON SELECTED CHAPTER SECTIONS

10.1 Temperature and the Zeroth Law of Thermodynamics

The zeroth law of thermodynamics (or the equilibrium law) can be stated as follows:

> *If bodies A and B are separately in thermal equilibrium with a third body, C, then A and B will be in thermal equilibrium with each other if placed in thermal contact.*

Two objects in thermal equilibrium with each other are at the same temperature.

10.2 Thermometers and Temperature Scales

The physical property used in a constant volume gas thermometer is the pressure variation with temperature of a fixed volume of gas. The temperature readings are nearly independent of the substance used in the thermometer.

The **triple point of water**, which is the single temperature and pressure at which water, water vapor, and ice can coexist in equilibrium, was chosen as a convenient and reproducible reference temperature for the Kelvin scale. It occurs at a temperature of 0.01°C and a pressure of 4.58 mm of mercury. The temperature at the triple point of water on the Kelvin scale has been assigned a value of 273.16 kelvins (K). Thus, the SI unit of temperature, the kelvin, is defined as 1/273.16 of the temperature of the triple point of water.

10.3 Thermal Expansion of Solids and Liquids

Liquids generally increase in volume with increasing temperature and have volume expansion coefficients about ten times greater than those of solids. Water is an exception to this rule; as the temperature increases from 0°C to 4°C, water contracts and thus its density increases. Above 4°C, water expands with increasing temperature. *The density of water reaches its maximum value (1 000 kg/m^3) four degrees above the freezing point.*

10.4 Macroscopic Description of an Ideal Gas

In an **ideal gas** the atoms or molecules move randomly, the individual particles do not exert long-range forces on each other, and each particle is considered to be a "point" mass. One **mole** of any substance is that quantity of material that contains a number of particles equal to the number of atoms in 12 grams of the isotope carbon-12. **One mole of any gas contains Avogadro's number of particles (N_A), and equal volumes of all gases at the same temperature and pressure contain the same number of particles.**

10.5 The Kinetic Theory of Gases

A microscopic **model of an ideal gas** is based on the following assumptions:

- **The number of molecules is large, and the average separation between them is large** compared with their dimensions. Therefore, the molecules occupy a negligible volume compared with the volume of the container.
- **The molecules obey Newton's laws of motion, but the individual molecules move in a random fashion**. By random fashion, we mean that the molecules move in all directions with equal probability and with various speeds. This distribution of velocities does not change in time, despite the collisions between molecules.
- **The molecules undergo elastic collisions with each other**. Thus, the molecules are considered to be without structure (that is, point masses), and in the collisions both kinetic energy and momentum are conserved.
- **The forces between molecules are negligible, except during a collision**. The forces between molecules are short-range, so that the only time the molecules interact with each other is during a collision.
- **The gas under consideration is a pure gas**. That is, all molecules are identical.
- **The molecules of the gas make perfectly elastic collisions with the walls of the container**. Hence, the wall will eject as many molecules as it absorbs, and the ejected molecules will have the same average kinetic energy as the absorbed molecules.

EQUATIONS AND CONCEPTS

The **Celsius temperature** T_C is related to the Kelvin temperature (sometimes called the absolute temperature) according to Equation 10.1. *The size of a degree on the Kelvin scale equals the size of a degree on the Celsius scale.*

$$T_C = T - 273.15 \qquad (10.1)$$

T_C = Celsius temperature

T = Kelvin temperature

Conversion between Fahrenheit and Celsius scales is based on the freezing and boiling point temperatures of water set for each scale.

$$T_F = \tfrac{9}{5}T_C + 32 \qquad (10.2a)$$

$$T_C = \tfrac{5}{9}(T_F - 32) \qquad (10.2b)$$

Freezing point = 0° C = 32°F

Boiling point = 100° C = 212°F

Thermal expansion results in a change in length, surface area, and volume. In each case, the **fractional change** is proportional to the change in temperature and a thermal coefficient characteristic of the particular type of material.

α = coefficient of linear expansion

$\gamma \cong 2\alpha$ = coefficient of area expansion

$\beta \cong 3\alpha$ = coefficient of volume expansion

Length expansion (solids)

$$\Delta L = \alpha L_0 \Delta T \qquad (10.4)$$

$$\frac{\Delta L}{L_0} = \alpha(\Delta T)$$

Area expansion (solids)

$$\Delta A = \gamma A_0 \Delta T \qquad (10.5)$$

$$\frac{\Delta A}{A_0} = \gamma(\Delta T)$$

Volume expansion (solids and liquids)

$$\Delta V = \beta V_0 \Delta T \qquad (10.6)$$

$$\frac{\Delta V}{V_0} = \beta(\Delta T)$$

The **equation of state of an ideal gas** is shown in two forms: expressed in terms of the number of moles (n) and the universal gas constant (R) as in Equation 10.8 or in terms of the number of molecules (N) and Boltzmann's constant (k_B) as in Equation 10.11. *The temperature T must always be expressed in kelvins.*

In terms of number of moles (n):

$$PV = nRT \qquad (10.8)$$

$$R = 8.31 \text{ J/mol} \cdot \text{K} = 0.0821 \text{ L} \cdot \text{atm/mol} \cdot \text{K}$$

In terms of number of molecules (N):

$$PV = Nk_BT \qquad (10.11)$$

$$k_B = \frac{R}{N_A} = 1.38 \times 10^{-23} \text{ J/K} \qquad (10.12)$$

The number of moles in a sample of an element or a compound can be stated in two ways:

The ratio of the mass of the sample to the atomic or molar mass characteristic of that particular material (Equation 10.7).

$$n = \frac{m}{\text{molar mass}} \qquad (10.7)$$

The number of atoms or molecules in the sample divided by Avogadro's number (Equation 10.13.)

$$n = \frac{N}{N_A} \qquad (10.10)$$

1 mole of a substance contains Avogadro's number of particles.

The following three laws are the result of experimental observations:

Boyle's law states that when a gas is maintained at constant temperature, the pressure is inversely proportional to the volume.

$$P \propto \left(\frac{1}{V}\right) \quad \text{(at constant temperature)}$$

Charles's law states that when a gas is maintained at constant pressure, volume is directly proportional to the absolute temperature.

$$P \propto \left(\frac{1}{V}\right) \quad \text{(at constant temperature)}$$

Gay-Lussac's law states that when the volume of a gas is held constant, the pressure is directly proportional to the absolute temperature.

$$P \propto T \quad \text{(at constant volume)}$$

The **pressure of an ideal gas** is proportional to the number of molecules per unit volume and to the average kinetic energy of the molecules.

$$P = \frac{2}{3}\left(\frac{N}{V}\right)\left(\tfrac{1}{2}m\overline{v^2}\right) \tag{10.13}$$

The **average translational kinetic energy** per molecule is directly proportional to the absolute temperature of the gas.

$$\tfrac{1}{2}m\overline{v^2} = \tfrac{3}{2}k_BT \tag{10.15}$$

The **total translational kinetic energy** of a sample of gas depends on the number of particles and the absolute temperature.

$$KE_{\text{total}} = N(\tfrac{1}{2}mv^2) = \tfrac{3}{2}Nk_BT = \tfrac{3}{2}nRT \tag{10.16}$$

The **root-mean-square (rms) speed** expression shows that, at a given temperature, lighter molecules move faster on the average than heavier ones.

$$v_{\text{rms}} = \sqrt{\overline{v^2}} = \sqrt{\frac{3k_BT}{m}} = \sqrt{\frac{3RT}{M}} \tag{10.18}$$

m = molecular mass

M = molar mass in kg/mole

REVIEW CHECKLIST

- Describe the operation of the constant-volume gas thermometer and how it is used to determine the Kelvin temperature scale. Convert between the various temperature scales, especially the conversion from degrees Celsius into kelvins, degrees Fahrenheit into kelvins, and degrees Celsius into degrees Fahrenheit.

- Define the linear expansion coefficient and volume expansion coefficient for an isotropic solid, and understand how to use these coefficients in practical situations involving expansion or contraction.

- Understand the assumptions made in developing the molecular model of an ideal gas; and apply the equation of state for an ideal gas to calculate pressure, volume, temperature, or number of moles.

- Define each of the following terms: molecular weight, mole, Avogadro's number, universal gas constant, and Boltzmann's constant.

SOLUTIONS TO SELECTED END-OF-CHAPTER PROBLEMS

3. The boiling point of liquid hydrogen is 20.3 K at atmospheric pressure. What is this temperature on (a) the Celsius scale and (b) the Fahrenheit scale?

Solution

(a) The scale divisions (degrees) on the Celsius and Kelvin temperature scales are the same size. The only difference in the two scales is that the zero point on the Kelvin scale is 273.15 divisions lower than the zero point on the Celsius scale. Thus, to convert from the Celsius scale to the Kelvin scale, it is only necessary to add 273.15 degrees to the Celsius temperature. That is: $T_K = T_C + 273.15$. If the boiling point of liquid hydrogen at atmospheric pressure is $T_K = 20.3$ K, the corresponding temperature on the Celsius scale is

$$T_C = T_K - 273.15 = 20.3 - 273.15 = -253°\text{C} \qquad \Diamond$$

(b) The Fahrenheit and Celsius temperature scales differ from each other in two respects. First, the Celsius scale division is almost double (actually 9/5 or 1.8 times) the size of the Fahrenheit scale division. Thus, in converting from the Celsius scale to the Fahrenheit scale, we must first multiply the number of Celsius divisions involved by 9/5 to get the number of Fahrenheit divisions involved. Secondly, the zero point on the Celsius scale is 32 Fahrenheit divisions higher than the zero point on the Fahrenheit scale. Therefore, we must add an additional 32 Fahrenheit divisions to account for the offset in the zero points. The conversion equation from Celsius temperatures to Fahrenheit temperatures is then $T_F = \frac{9}{5}T_C + 32$. On the Fahrenheit scale, the boiling point of liquid hydrogen at atmospheric pressure is

$$T_F = \frac{9}{5}T_C + 32 = \frac{9}{5}(-253) + 32 = -423°\text{F} \qquad \Diamond$$

---

10. Temperature differences on the Rankine scale are identical to differences on the Fahrenheit scale, but absolute zero is given as 0°R. (a) Find a relationship converting the temperatures T_F of the Fahrenheit scale to the corresponding temperatures T_R of the Rankine scale. (b) Find a second relationship converting temperatures T_R of the Rankine scale to the temperatures T_K of the Kelvin scale.

Solution

(a) Since the difference between two temperatures is the same (i.e., includes the same number of divisions) on the Rankine and Fahrenheit temperature scales, the Rankine and Fahrenheit scale divisions must be the same size. Thus, the two temperature scales differ only in the location of their zero points. The zero point on the Rankine scale is at absolute zero (or at –273.15°C). The Fahrenheit reading at this point is $T_F = \frac{9}{5}T_C + 32 = \frac{9}{5}(-273.15) + 32 = -459.67°F$. The zero point on the Rankine scale is therefore 459.67 divisions lower than the zero point on the Fahrenheit scale, and it is necessary to add this number of divisions to the Fahrenheit temperature to obtain the corresponding Rankine temperature. The desired relationship is then $T_R = T_F + 459.67°$. ◊

(b) To find the relation between a temperature T_R on the Rankine scale and the corresponding temperature T_K on the Kelvin scale, we start with T_R and convert it to the Kelvin scale using the result from part (a) above and the results obtained on the previous page in problem 3. First converting to the Fahrenheit scale, we have $T_F = T_R - 459.67$. Then, converting to the Celsius scale gives

$$T_C = \frac{5}{9}(T_F - 32) = \frac{5}{9}(T_R - 459.67 - 32) = \frac{5}{9}(T_R - 491.67)$$

Finally, converting to the Kelvin scale, we find

$$T_K = T_C + 273.15 = \frac{5}{9}T_R - \frac{5}{9}(491.67) + 273.15 = \frac{5}{9}T_R - \cancel{273.15} + \cancel{273.15}$$

and the desired conversion equation is

$$T_K = \frac{5}{9}T_R$$ ◊

13. A pair of eyeglass frames are made of epoxy plastic (coefficient of linear expansion $= 1.30 \times 10^{-4}\ °\text{C}^{-1}$). At room temperature (20.0°C), the frames have circular lens holes 2.20 cm in radius. To what temperature must the frames be heated if lenses 2.21 cm in radius are to be inserted into them?

Solution

There are two approaches one could take in the solution of this problem. One is to realize that, at 20.0°C, the epoxy material forming the circumference of one of the holes in the frames is shorter than the circumference of the lens. Then, the problem is to find the temperature to which the epoxy frames should be heated so the original circumference of the hole, $C_0 = 2\pi r_0 = 2\pi(2.20\text{ cm})$, will have expanded to match the circumference, $C = 2\pi r = 2\pi(2.21\text{ cm})$, of the lens. Taking this approach and using the linear expansion equation, we have $C = C_0[1+\alpha(\Delta T)]$, or

$$2\pi(2.21\text{ cm}) = 2\pi(2.20\text{ cm})\left[1+\left(1.30\times10^{-4}\ °\text{C}^{-1}\right)(T-20.0°\text{C})\right] \qquad \textbf{[1]}$$

giving

$$T - 20.0°\text{C} = \frac{1}{1.30\times10^{-4}\ °\text{C}^{-1}}\cdot\left[\frac{\cancel{2\pi}(2.21\text{ cm})}{\cancel{2\pi}(2.20\text{ cm})}-1\right] = 35.0°\text{C}$$

and the desired temperature is

$$T = 35.0°\text{C} + 20.0°\text{C} = 55.0°\text{C} \qquad \Diamond$$

The other way one could think of this problem is to consider the expansion of the radius of the hole, realizing that linear dimensions of a hole or void in a surrounding material expand exactly the same as they would if filled with the surrounding material. That is: When the epoxy material is heated, the circular hole in that material expands exactly the same as a circular disk of epoxy material having the same radius as the hole (perhaps the disk that was cut away to create the hole) would expand. If we take this approach, we are computing the temperature to which the epoxy material should be heated so the original radius of the hole will expand to match the radius of the lens. Then, we would write the linear expansion equation (using the expansion coefficient of the surrounding epoxy) as $r = r_0[1+\alpha(\Delta T)]$, or

$$(2.21\text{ cm}) = (2.20\text{ cm})\left[1+\left(1.30\times10^{-4}\ °\text{C}^{-1}\right)(T-20.0°\text{C})\right]$$

Observe that this is the same as equation [1] above with the factors of 2π canceled. Thus, the two approaches are fully equivalent.

17. Lead has a density of $11.3\times10^3\ \text{kg/m}^3$ at $0°\text{C}$. (a) What is the density of lead at $90°\text{C}$? (b) Based on your answer to part (a), now consider a situation in which you plan to invest in a gold bar. Would you be better off buying it on a warm day? Explain.

Solution

(a) As the temperature of a material rises, the mass of the material present does not change; however, its volume increases. Therefore, the density (the mass per unit volume) will decrease with increasing temperature. The original density will be $\rho_0 = m/V_0$, where m is the mass of material present and V_0 is the volume occupied at the original temperature.

At the new temperature, the density is $\rho = m/V$, where the new volume is $V = V_0 + \Delta V$. The change in volume ΔV which occurs with change in temperature ΔT is given by the volume expansion equation as $\Delta V = \beta V_0(\Delta T)$, so $V = V_0 + \beta V_0(\Delta T) = V_0[1+\beta(\Delta T)]$ and

$$\rho = \frac{m}{V_0[1+\beta(\Delta T)]} = \frac{\rho_0}{1+\beta(\Delta T)}$$

With $\beta_{\text{lead}} = 3\alpha_{\text{lead}}$, $\Delta T = 90.0°\text{C} - 0°\text{C} = 90.0°\text{C}$, and $\rho_{0,\text{lead}} = 11.3\times10^3\ \text{kg/m}^3$, the density of lead at $90.0°\text{C}$ is found to be

$$\rho_{\text{lead}} = \frac{11.3\times10^3\ \text{kg/m}^3}{1+3\left(29\times10^{-6}\ °\text{C}^{-1}\right)(90.0°\text{C})} = 11.2\times10^3\ \text{kg/m}^3 \qquad \Diamond$$

(b) You would not be better off buying the gold bar on a warm day. While the density of gold would be less on a warm day, the mass of the bar would be the same and that is what you are paying for. As the temperature rises, the volume increases, causing the density to decrease, but the mass remains constant.

---

25. The average coefficient of volume expansion for carbon tetrachloride is $5.81\times10^{-4}\ (°\text{C})^{-1}$. If a 50.0-gal steel container is filled completely with carbon tetrachloride when the temperature is $10.0°\text{C}$, how much will spill over when the temperature rises to $30.0°\text{C}$?

Solution

As the temperature rises, both the carbon tetrachloride and the steel container will expand. However, the volume expansion coefficient for carbon tetrachloride, $\beta_{\text{C(Cl)}_4} = 5.81\times10^{-4}\ (°\text{C})^{-1}$, is greater than that of steel, $\beta_{\text{steel}} = 3\alpha_{\text{steel}} = 33.0\times10^{-6}\ (°\text{C})^{-1}$. Thus, the carbon tetrachloride expands faster than the steel container, and since the container was completely filled at the original temperature, spillage will occur as the temperature rises.

The amount of spillage that occurs is the amount by which the expansion of the carbon tetrachloride volume exceeds the expansion of the steel container's volume. That is

$$V_{\text{spillage}} = \Delta V_{\text{C(Cl}_4)} - \Delta V_{\substack{\text{steel}\\\text{container}}} = \beta_{\text{C(Cl}_4)} V_0(\Delta T) - \beta_{\text{steel}} V_0(\Delta T) = V_0\left(\beta_{\text{C(Cl}_4)} - \beta_{\text{steel}}\right)(\Delta T)$$

Note that the original volume, $V_0 = 50.0$ gal, and the change in temperature, $\Delta T = +20.0°\text{C}$, are the same for both the carbon tetrachloride and the steel container. Also, note that we are treating the steel container as though it were solid steel, rather than a steel surface around a cavity. Please see the solution of Problem 13 earlier in this chapter for a discussion of this concept as it relates to linear expansion.

We then have

$$V_{\text{spillage}} = (50.0 \text{ gal})\left[5.81\times10^{-4}\ (°\text{C})^{-1} - 33.0\times10^{-6}\ (°\text{C})^{-1}\right](20.0°\text{C}) = 0.548 \text{ gal} \qquad \Diamond$$

---

33. Gas is confined in a tank at a pressure of 11.0 atm and a temperature of 25.0°C. If two-thirds of the gas is withdrawn and the temperature is raised to 75.0°C, what is the new pressure of the gas remaining in the tank?

Solution

We shall assume the gas approximates an ideal gas and apply the ideal gas law, $PV = nRT$. Recall that when using the ideal gas law, absolute temperatures are required. On the Kelvin scale, the initial and final temperatures of this gas are

$$T_i = 25.0 + 273.15 = 298 \text{ K}$$

and

$$T_f = 75.0 + 273.15 = 348 \text{ K}$$

If, as the temperature is being raised, two-thirds of the gas is withdrawn from the tank, the final number of moles in the tank is one-third the initial number of moles present, or $n_f = \frac{1}{3}n_i$. Using the ideal gas law, with the volume constant, to form a ratio of the initial and final conditions gives

$$\frac{P_f \cancel{V_f}}{P_i \cancel{V_i}} = \frac{n_f \cancel{R} T_f}{n_i \cancel{R} T_i}$$

or

$$P_f = P_i\left(\frac{n_f}{n_i}\right)\left(\frac{T_f}{T_i}\right) = (11.0 \text{ atm})\left(\frac{1}{3}\right)\left(\frac{348 \text{ K}}{298 \text{ K}}\right) = 4.28 \text{ atm} \qquad \Diamond$$

---

37. An air bubble has a volume of 1.50 cm^3 when it is released by a submarine 100 m below the surface of a lake. What is the volume of the bubble when it reaches the surface? Assume the temperature and the number of air molecules in the bubble remain constant during its ascent.

Solution

If the bubble rises to the surface of the lake rapidly enough, there will be insufficient time for the temperature of the trapped air to change significantly. Also, if the air bubble does not break up as it rises, the number of moles of air contained in it is constant. The pressure exerted on the trapped air by the water will decrease as the bubble gets closer to the surface of the lake. This decreased pressure will allow the bubble to expand in volume.

Using the ideal gas law, with both the temperature and number of moles constant, we find that

$$\frac{P_f V_f}{P_i V_i} = \frac{n_f \cancel{RT_f}}{n_i \cancel{RT_i}} = 1$$

or

$$V_f = \left(\frac{P_i}{P_f}\right) V_i$$

With a pressure of $P_0 = 1 \text{ atm} = 1.013 \times 10^5 \text{ Pa}$ at the surface, the pressure at a depth of 100 m below the surface of the water is

$$P_i = P_0 + \rho_{\text{water}} gh = 1.013 \times 10^5 \text{ Pa} + \left(1.00 \times 10^3 \text{ kg/m}^3\right)\left(9.80 \text{ m/s}^2\right)\left(100 \text{ m}\right)$$

or

$$P_i = 1.08 \times 10^6 \text{ Pa}$$

Then, with $P_f = 1 \text{ atm} = 1.013 \times 10^5 \text{ Pa}$, we find the volume of the bubble when it reaches the surface of the lake to be

$$V_f = \left(\frac{1.08 \times 10^6 \text{ Pa}}{1.013 \times 10^5 \text{ Pa}}\right)\left(1.50 \text{ cm}^3\right) = 16.0 \text{ cm}^3 \qquad \Diamond$$

41. Use Avogadro's number to find the mass of a helium atom.

Solution

One mole of any substance contains Avogadro's number N_A of molecules and has a mass equal to the molar mass or molecular weight M of that substance. Thus, if m_{molecule} is the mass of a single molecule of the substance, we have

$$N_A m_{\text{molecule}} = M$$

or

$$m_{\text{molecule}} = \frac{M}{N_A}$$

Helium, an inert gas, is a monatomic substance. This means that each molecule of helium contains a single helium atom, so the mass of a helium molecule and that of a helium atom are identical. Since the molecular weight of helium is $M_{\text{He}} = 4.00\ \text{g/mol}$, and Avogadro's number is $N_A = 6.02 \times 10^{23}$ molecules/mol, we have for helium

$$m_{\text{molecule}} = \frac{M_{\text{He}}}{N_A} = \frac{4.00\ \text{g/mol}}{6.02 \times 10^{23}\ \text{molecules/mol}} = 6.64 \times 10^{-24}\ \text{g/molecule}$$

and

$$m_{\text{atom}} = m_{\text{molecule}} = 6.64 \times 10^{-24}\ \text{g/atom} \qquad \Diamond$$

---

44. A 7.00-L vessel contains 3.50 moles of ideal gas at a pressure of 1.60×10^6 Pa. Find (a) the temperature of the gas and (b) the average kinetic energy of a gas molecule in the vessel. (c) What additional information would you need if you were asked to find the average speed of a gas molecule?

Solution

(a) The volume the gas occupies is

$$V = 7.00\ \text{L} = (7.00\ \cancel{\text{L}})\left(\frac{10^3\ \cancel{\text{cm}^3}}{1\ \cancel{\text{L}}}\right)\left(\frac{1\ \text{m}^3}{10^6\ \cancel{\text{cm}^3}}\right) = 7.00 \times 10^{-3}\ \text{m}^3$$

The ideal gas law, $PV = nRT$, then gives the absolute temperature as

$$T = \frac{PV}{nR} = \frac{(1.60 \times 10^6\ \text{Pa})(7.00 \times 10^{-3}\ \text{m}^3)}{(3.50\ \text{mol})(8.31\ \text{J/mol} \cdot \text{K})} = 385\ \text{K} \qquad \Diamond$$

(b) From kinetic theory, the average kinetic energy of a molecule in an ideal gas (or any gas that approximates an ideal gas) is

$$\overline{KE}_{\text{molecule}} = \frac{1}{2} m\overline{v^2} = \frac{3}{2} k_B T$$

where k_B is Boltzmann's constant and T is the absolute temperature of the gas. Thus, for the given gas, we have

$$\overline{KE}_{\text{molecule}} = \frac{3}{2}\left(1.38\times10^{-23}\ \text{J/K}\right)\left(385\ \text{K}\right) = 7.97\times10^{-21}\ \text{J}$$ ◊

(c) The average molecular speed (actually the rms or root-mean-square speed) can be found from

$$\overline{KE}_{\text{molecule}} = m\overline{v^2}/2 \qquad \text{as} \qquad v_{\text{rms}} = \sqrt{\overline{v^2}} = \sqrt{2\overline{KE}_{\text{molecule}}/m}$$

To carry out this computation, one needs the mass m of a single gas molecule, which in turn requires a knowledge of the molecular weight of the gas. (See the solution of Problem 41 earlier in this chapter.) ◊

---

53. Long-term space missions require reclamation of the oxygen in the carbon dioxide exhaled by the crew. In one method of reclamation, 1.00 mol of carbon dioxide produces 1.00 mol of oxygen, with 1.00 mol of methane as a by-product. The methane is stored in a tank under pressure and is available to control the attitude of the spacecraft by controlled venting. A single astronaut exhales 1.09 kg of carbon dioxide each day. If the methane generated in the recycling of three astronauts' respiration during one week of flight is stored in an originally empty 150-L tank at −45.0°C, what is the final pressure in the tank?

Solution

During a one-week flight, the three astronauts will generate $3(1.09\ \text{kg}) = 3.27\ \text{kg}$ of carbon dioxide each day, or a total of $7(3.27\ \text{kg}) = 22.9\ \text{kg}$ for the week. Since the molecular weight of carbon dioxide is $M = 44.0\ \text{g/mol} = 44.0\times10^{-3}\ \text{kg/mol}$, the number of moles of carbon dioxide generated is

$$n = \frac{22.9\ \text{kg}}{44.0\times10^{-3}\ \text{kg/mol}} = 520\ \text{mol}$$

From this, a total of $n = 520\ \text{mol}$ of methane will be produced as a by-product of the reclamation process. When stored at an absolute temperature of

$$T = T_C + 273.15 = -45.0 + 273.15 = 228\ \text{K}$$

in a tank with volume $V = 150\ \text{L} = 150\times10^3\ \text{cm}^3 = 150\times10^{-3}\ \text{m}^3$, the methane will exert a pressure of

$$P = \frac{nRT}{V} = \frac{(520)(8.31\ \text{J/mol}\cdot\text{K})(228\ \text{K})}{150\times10^{-3}\ \text{m}^3} = 6.57\times10^6\ \text{Pa} = 6.57\ \text{MPa}$$ ◊

---

59. Two concrete spans of a 250-m-long bridge are placed end to end so that no room is allowed for expansion (Fig. P10.59a). If the temperature increases by 20.0°C, what is the height y to which the spans rise when they buckle (Fig. P10.59b)?

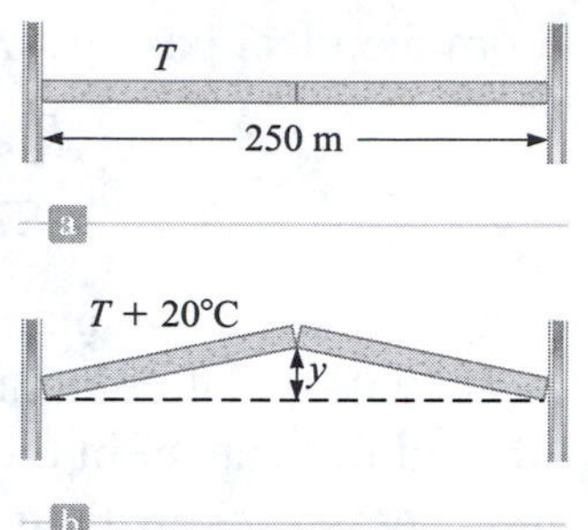

Figure P10.59

Solution

Each span of this two-span concrete bridge has an initial length of $L_0 = 125$ m and a coefficient of linear expansion of $\alpha_{\text{concrete}} = 12\times10^{-6}\ (°\text{C})^{-1}$ (see Table 10.1 in the textbook). As the temperature of the bridge increases by 20.0°C, each span undergoes an expansion of the amount

$$\Delta L = L_0\alpha_{\text{concrete}}(\Delta T) = (125\text{ m})(12\times10^{-6}\ °\text{C}^{-1})(20.0\ °\text{C}) = 3.0\times10^{-2}\text{ m} = 3.0\text{ cm}$$

After expansion, each span has an overall length of $L = 125\text{ m} + 3.0\times10^{-2}\text{ m}$. As shown in Figure P10.59b, this increased length of the span forms the hypotenuse of a right triangle having the original 125-m length as its base. The altitude of this triangle is the vertical displacement y produced by the buckling of the bridge.

We use the Pythagorean theorem to find the magnitude of this vertical displacement as

$$y = \sqrt{L^2 - L_0^2} = \sqrt{[(125+0.030)\text{ m}]^2 - (125\text{ m})^2}$$

which yields

$$y = 2.7\text{ m} \qquad (\text{or } \approx 8.9\text{ ft})$$ ◊

---

64. Two small containers, each with a volume of 100 cm^3, contain helium gas at 0°C and 1.00 atm pressure. The two containers are joined by a small open tube of negligible volume, allowing gas to flow from one container to the other. What common pressure will exist in the two containers if the temperature of one container is raised to 100°C while the other container is kept at 0°C?

Solution

We shall call the container that is kept at 0°C container 1 and the one whose temperature is raised to 100°C will be container 2.

As the temperature of container 2 is raised, gas may flow from one container to the other, but no gas escapes. Thus, at the end, the total number of moles of gas in the joined containers is the same as at the beginning, or

$$n_{1f} + n_{2f} = n_{1i} + n_{2i}$$

From the ideal gas law, $PV = nRT$, we have $n = PV/RT$ and our previous result becomes

$$\frac{P_{1f}V_{1f}}{RT_{1f}} + \frac{P_{2f}V_{2f}}{RT_{2f}} = \frac{P_{1i}V_{1i}}{RT_{1i}} + \frac{P_{2i}V_{2i}}{RT_{2i}}$$

Since the two containers have equal initial volumes, and ignoring the very slight change in the volume of container 2 as its temperature is raised, we have $V_{1i} = V_{2i} = V_{1f} = V_{2f}$ and all volumes cancel in the above equation. Also, the two containers are joined, maintaining equal pressures in them at all times. Thus, $P_{1f} = P_{2f} = P_f$ and $P_{1i} = P_{2i} = 1.00$ atm. With these values, our previous result becomes

$$P_f\left[\frac{\cancel{V_{1f}}}{\cancel{R}T_{1f}} + \frac{\cancel{V_{2f}}}{\cancel{R}T_{2f}}\right] = (1.00\text{ atm})\left[\frac{\cancel{V_{1i}}}{\cancel{R}T_{1i}} + \frac{\cancel{V_{2i}}}{\cancel{R}T_{2i}}\right]$$

or

$$P_f = (1.00\text{ atm})\left(\frac{T_{2i} + T_{1i}}{T_{1i}T_{2i}}\right)\left(\frac{T_{1f}T_{2f}}{T_{2f} + T_{1f}}\right)$$

Finally, we recognize that the absolute temperatures are $T_{1i} = T_{2i} = T_{1f} = 0°\text{C} = 273\text{ K}$ and $T_{2f} = 100°\text{ C} = 373\text{ K}$, so we have

$$P_f = (1.00\text{ atm})\left[\frac{273\text{ K} + 273\text{ K}}{(273\text{ K})\cancel{(273\text{ K})}}\right]\left[\frac{\cancel{(273\text{ K})}(373\text{ K})}{373\text{ K} + 273\text{ K}}\right] = 1.15\text{ atm} \qquad \Diamond$$

Observe that, in order to solve this problem, it was not necessary to know the actual volumes of the containers, only that each was very large in comparison to any increase in the volume of container 2 as it was heated. Also, it was not necessary to know the specific gas in the container, only that it behaved like an ideal gas.

11

Energy in Thermal Processes

NOTES ON SELECTED CHAPTER SECTIONS

11.1 Heat and Internal Energy

When two systems at different temperatures are placed in thermal contact, energy is transferred by heat from the warmer to the cooler object until they reach a common temperature (i.e., when they are in thermal equilibrium with each other). Heat is a mechanism by which energy (measured in calories or joules) is transferred between a system and its environment due to a temperature difference.

The **mechanical equivalent of heat**, first measured by Joule, is given by $1 \text{ cal} \equiv 4.186 \text{ J}$. This is the definition of the calorie.

11.2 Specific Heat

The **specific heat**, c, of a substance is the energy per unit mass per degree Celsius temperature increase for that substance.

11.3 Calorimetry

Calorimetry is a procedure carried out in an isolated system (usually a substance of unknown specific heat and water) in which the transfer of thermal energy occurs. A negligible amount of mechanical work is done in the process. The law of conservation of energy in a calorimeter requires that the energy that leaves the warmer substance (of unknown specific heat) equals the energy that enters the water.

11.4 Latent Heat and Phase Change

A substance usually undergoes a change in temperature when energy is transferred by heat between it and its surroundings. There are situations, however, in which the transfer of energy does not result in a change in temperature. This is the case whenever the substance undergoes a **phase change**. Some common phase changes are solid to liquid, liquid to gas, and a change in the crystalline structure of a solid. *Every phase change involves a change in internal energy without an accompanying change in temperature.*

The **latent heat of fusion** is a parameter used to characterize a solid-to-liquid phase change; the **latent heat of vaporization** characterizes the liquid-to-gas phase change.

11.5 Energy Transfer

There are three basic processes of thermal energy transfer: conduction, convection, and radiation.

Conduction is an energy transfer process that occurs in a substance when there is a temperature gradient across the substance. That is, conduction of energy occurs only when the temperature of the substance is **not** uniform. For example, if there is a temperature difference between the two ends of a metal rod, energy will flow from the hot end to the colder end. The rate of flow of heat along the rod is proportional to the cross-sectional area of the rod, the temperature gradient, and k, the thermal conductivity of the material of which the rod is made.

Convection is a process of energy transfer by the motion of material, such as the mixing of hot and cold fluids. Convection heating is used in conventional hot-air and hot-water heating systems. Convection currents produce changes in weather conditions when warm and cold air masses mix in the atmosphere.

Radiation is energy transfer by the emission or absorption of electromagnetic waves.

EQUATIONS AND CONCEPTS

The **mechanical equivalent of heat** was first measured by Joule. Equation 11.1 is the definition of the calorie as an energy unit.

$$1 \text{ cal} \equiv 4.186 \text{ J} \quad (11.1)$$

The **specific heat**, c, of a substance is defined by Equation 11.2. As seen in Equation 11.3, c determines the quantity of energy that must be transferred in order to change the temperature of a mass m of a specific material by an amount ΔT. *Specific heat is characteristic of a* ***specific type*** *of material.*

$$c \equiv \frac{Q}{m\Delta T} \quad (11.2)$$

$$\text{or } Q = mc\Delta T \quad (11.3)$$

SI units of c = J/kg · °C

Calorimetry, a technique to measure specific heat, is based on conservation of energy in an isolated system. *The negative sign is required so that both sides of Equation 11.4 will be positive.*

$$Q_{\text{cold}} = -Q_{\text{hot}} \quad (11.4)$$

Q_{cold} is a positive quantity.

Q_{hot} is a negative quantity.

Latent heat (L) Is a thermal property of a substance that determines the quantity of energy required to cause a mass m of that substance to undergo a phase change. The latent heat of fusion, L_f, is used when the phase change is from solid to liquid (or liquid to solid). The latent

$$Q = \pm mL \quad (11.6)$$

Use + when energy enters a system.

Use – when energy leaves a system.

L is expressed in units of J/kg.

heat of vaporization, L_v, is used when the phase change is from liquid to gas (or gas to liquid). *A phase change occurs at constant temperature and is accompanied by a change in internal energy.*

Energy transfer by conduction through a slab or rod of material is directly proportional to the temperature difference between the hot and cold faces and to the cross-sectional area of the slab and indirectly proportional to the thickness of the slab (or the length of the rod). *The thermal conductivity parameter (k) is a characteristic of a particular material.*

$$P = kA\left(\frac{T_h - T_c}{L}\right) \tag{11.7}$$

T_h

A

Energy transfer for $T_h > T_c$

T_c

Δx

Energy transfer through a compound slab is accomplished by a summation over all layers of the slab. For this calculation, T_h and T_c are the temperatures of the outer extremities of the slab. The thicknesses of the layers of the slab are $L_1, L_2, L_3, \ldots$, and $k_1, k_2, k_3, \ldots$ are respective thermal conductivities. The rate of energy transfer can also be expressed in cal/s and Btu/h.

$$\frac{Q}{\Delta t} = \frac{A(T_h - T_c)}{\sum_i L_i / k_i} \tag{11.8}$$

$$\text{or } \frac{Q}{\Delta t} = \frac{A(T_h - T_c)}{\sum_i R_i} \tag{11.9}$$

$$\text{where } R_i = \frac{L_i}{k_i}$$

Stefan's law expresses the **rate of energy transfer by radiation** (radiated power). *The radiated power is proportional to the fourth power of the absolute temperature.*

$$P = \sigma A e T^4 \tag{11.10}$$

$$\sigma = 5.669\,6 \times 10^{-8}\ \text{W/m}^2\cdot\text{K}^4$$

T = surface temperature in kelvins

e = emissivity (values from 0 to 1)

Net radiated power by an object at temperature T in an environment at temperature T_0 is the difference between radiated and absorbed power. *At thermal equilibrium ($T = T_0$), an object radiates and absorbs energy at the same rate and the temperature of the object remains constant.*

$$P_{\text{net}} = \sigma A e\left(T^4 - T_0^4\right) \tag{11.11}$$

SUGGESTIONS, SKILLS, AND STRATEGIES

If you are having difficulty with calorimetry problems, one or more of the following factors should be considered:

1. Be sure your units are consistent throughout. That is, if you are using specific heats measured in J/kg · °C, be sure that masses are in kilograms and temperatures are in Celsius units throughout.

2. Losses and gains in energy are found by using $Q = mc\Delta T$ only for those intervals in which no phase changes occur. Likewise, the equations $Q = \pm m L_f$ and $Q = \pm m L_v$ are to be used only when phase changes are taking place.

3. Often sign errors occur when applying the basic calorimetry equation ($Q_{\text{cold}} = -Q_{\text{hot}}$). Remember to include the negative sign in the equation; and remember that **ΔT is always the final temperature minus the initial temperature**.

4. When three or more substances exchange thermal energy in a calorimeter, Equation 11.5 ($\Sigma Q_k = 0$) should be used to solve for an unknown quantity (temperature, mass, or specific heat). In this equation Q_k refers to the *k*th substance and is given by Equation 11.3.

REVIEW CHECKLIST

- Define and discuss the calorie, Btu, specific heat, and latent heat. Convert among calories, Btu's, and joules.

- Use equations for specific heat, latent heat, temperature change, and energy gain (loss) to solve calorimetry problems.

- Discuss the possible mechanisms which can give rise to energy transfer between a system and its surroundings (i.e., conduction, convection, and radiation), and give a realistic example of each transfer mechanism.

- Apply the basic law of thermal conduction, and Stefan's law for energy transfer by radiation.

SOLUTIONS TO SELECTED END-OF-CHAPTER PROBLEMS

5. A 3.00-g copper coin at 25.0°C drops 50.0 m to the ground. (a) Assuming 60.0% of the change in gravitational potential energy of the coin-Earth system goes into increasing the internal energy of the coin, determine the coin's final temperature. (b) Does the result depend on the mass of the coin? Explain.

Solution

(a) When a coin of mass *m* drops from height *h* to the ground, the gravitational potential energy the coin-Earth system gives up is $|\Delta PE_g| = mgh$. If 60.0% of this is converted into internal energy of the coin, the increase in the internal energy of the coin is $Q = +0.600mgh$. Then, from the definition of specific heat, $Q = mc\Delta T$, the change in temperature the coin will experience is

$$\Delta T = \frac{Q}{mc_{\text{Cu}}} = \frac{+0.600\,\cancel{m}gh}{\cancel{m}c_{\text{Cu}}} = \frac{+0.600gh}{c_{\text{Cu}}}$$

Using the given height and the specific heat of copper from Table 11.1 of the textbook, we find

$$\Delta T = \frac{+0.600\left(9.80\ \text{m/s}^2\right)(50.0\ \text{m})}{387\ \text{J/kg}\cdot{}^\circ\text{C}} = +0.760^\circ\text{C}$$

and the final temperature of the coin will be

$$T_f = T_i + \Delta T = 25.0^\circ\text{C} + 0.760^\circ\text{C} = 25.8^\circ\text{C}$$ ◊

(b) As can be observed in the above calculation, the mass of the coin cancels out and therefore the final result does not depend on the mass of the coin. All copper coins falling from the given height will experience the same change in temperature. The reason for this is that both the loss of gravitational potential energy and the additional internal energy needed to produce a change ΔT in temperature are proportional to the mass of the coin. Thus, as the mass changes, both quantities change by the same factor and the ratio which determines the change in temperature is unaltered. ◊

11. A 200-g aluminum cup contains 800 g of water in thermal equilibrium with the cup at 80°C. The combination of cup and water is cooled uniformly so that the temperature decreases by 1.5°C per minute. At what rate is energy being removed? Express your answer in watts.

Solution

Note that the water in the cup will be liquid, while the aluminum of the cup will be solid, and neither the water nor the aluminum will undergo a change of phase for temperature decreases of less than 80°C. Consider a one-minute time interval during which the temperature of the system consisting of the 200-g aluminum cup and 800 g of water undergoes a change in temperature of $\Delta T = -1.5°\text{C}$. The internal energy that must be removed from the system to accomplish this change in temperature is

$$|Q| = m_{\text{cup}} c_{\text{Al}} |\Delta T| + m_{\text{water}} c_{\substack{\text{liquid} \\ \text{water}}} |\Delta T|$$

and using the values of specific heat from Table 11.1 of the textbook,

$$|Q| = (0.200 \text{ kg})(900 \text{ J/kg}\cdot°\text{C})(1.5°\text{C}) + (0.800 \text{ kg})(4\,186 \text{ J/kg}\cdot°\text{C})(1.5°\text{C})$$

yielding

$$|Q| = 5.3 \times 10^3 \text{ J}$$

The rate at which energy is being removed is then

$$P = \frac{|Q|}{\Delta t} = \frac{5.3 \times 10^3 \text{ J}}{1.0 \text{ min}}\left(\frac{1 \text{ min}}{60 \text{ s}}\right) = 88 \text{ J/s} = 88 \text{ W}$$ ◊

15. What mass of water at 25.0°C must be allowed to come to thermal equilibrium with a 1.85-kg cube of aluminum initially at 1.50×10^2 °C to lower the temperature of the aluminum to 65.0°C? Assume any water turned to steam subsequently recondenses.

Solution

We assume that the water-aluminum system is thermally isolated from the environment, so no energy is transferred between this system and the environment. Then, any energy given up by one part of the system must be absorbed by the other part of the system, with the total energy content of the system remaining constant. This restatement of the principle of conservation of energy is summarized by the equation $Q_{\text{cold}} = -Q_{\text{hot}}$.

We are also assuming that any steam produced by dropping the hot aluminum into the water recondenses before the final temperature of the system is reached. Without any permanent change of state, the energy gained or lost by any part of the system is related to the change in temperature of that part by $Q = mc\Delta T$, where c is the specific heat of the material making up that part of the system. Using this in the conservation of energy equation from above gives

$$m_{\text{water}}c_{\text{water}}\left(T_f - T_{i,\text{water}}\right) = -m_{\text{Al}}c_{\text{Al}}\left(T_f - T_{i,\text{Al}}\right)$$

so the mass of water needed to achieve the desired final temperature is

$$m_{\text{water}} = \frac{-m_{\text{Al}}c_{\text{Al}}\left(T_f - T_{i,\text{Al}}\right)}{c_{\text{water}}\left(T_f - T_{i,\text{water}}\right)} = \frac{-(1.85\text{ kg})(900\text{ J/kg}\cdot{}^\circ\text{C})(65.0^\circ\text{C} - 150^\circ\text{C})}{(4\,186\text{ J/kg}\cdot{}^\circ\text{C})(65.0^\circ\text{C} - 25.0^\circ\text{C})}$$

yielding

$$m_{\text{water}} = 0.845\text{ kg}$$ ◊

---

24. An unknown substance has a mass of 0.125 kg and an initial temperature of 95.0°C. The substance is then dropped into a calorimeter made of aluminum containing 0.285 kg of water initially at 25.0°C. The mass of the aluminum container is 0.150 kg, and the temperature of the calorimeter increases to a final equilibrium temperature of 32.0°C. Assuming no thermal energy is transferred to the environment, calculate the specific heat of the unknown substance.

Solution

If no thermal energy is transferred to or from the environment, any energy gained by the initially cooler materials must equal the energy given up by the initially warmer materials as the system approaches the equilibrium temperature. That is, $Q_{\text{cold}} = -Q_{\text{hot}}$, or

$$m_{\text{Al}}c_{\text{Al}}\left(T_f - T_{i,\text{Al}}\right) + m_{\text{water}}c_{\text{water}}\left(T_f - T_{i,\text{water}}\right) = -m_x c_x\left(T_f - T_{i,x}\right)$$

where the subscript Al refers to the aluminum calorimeter cup and the subscript x refers to the unknown substance. Observe that $T_{i,\text{Al}} = T_{i,\text{water}} = T_{\text{cold}} = 25.0^\circ\text{C}$. The specific heat of the unknown material is then

$$c_x = \frac{\left[m_{\text{Al}}c_{\text{Al}} + m_{\text{water}}c_{\text{water}}\right]\left(T_f - T_{\text{cold}}\right)}{-m_x\left(T_f - T_{i,x}\right)}$$

or

$$c_x = \frac{\left[(0.150\text{ kg})(900\text{ J/kg}\cdot{}^\circ\text{C}) + (0.285\text{ kg})(4\,186\text{ J/kg}\cdot{}^\circ\text{C})\right](32.0^\circ\text{C} - 25.0^\circ\text{C})}{-(0.125\text{ kg})(32.0^\circ\text{C} - 95.0^\circ\text{C})}$$

which yields

$$c_x = 1.18\times10^3\text{ J/kg}\cdot{}^\circ\text{C}$$ ◊

---

31. A 40-g block of ice is cooled to $-78°C$ and is then added to 560 g of water in an 80-g copper calorimeter at a temperature of 25°C. Determine the final temperature of the system consisting of the ice, water, and calorimeter. (If not all the ice melts, determine how much ice is left.) Remember that the ice must first warm to 0°C, melt, and then continue warming as water. The specific heat of ice is $0.500\ \text{cal/g}\cdot°\text{C} = 2\,090\ \text{J/kg}\cdot°\text{C}$.

Solution

The recommended first step in problems similar to this is to determine whether all of the ice will melt. To do this, we compute the minimum energy that must be added to the ice to melt all of it. This is $Q_{\text{melt}} = Q_{\substack{\text{warm}\\ \text{ice to 0°C}}} + Q_{\substack{\text{ice to}\\ \text{liquid at 0°C}}}$, or

$$Q_{\text{melt}} = m_{\text{ice}}c_{\text{ice}}\left(0°\text{C} - T_{i,\text{ice}}\right) + m_{\text{ice}}L_{f,\text{water}}$$

$$= (0.040\ \text{kg})\left[(2\,090\ \text{J/kg}\cdot°\text{C})(78°\text{C}) + 3.33\times10^{5}\ \text{J/kg}\right] = 2.0\times10^{4}\ \text{J}$$

The energy the warmer materials can give up before their temperature reaches 0°C is

$$Q_{\text{available}} = -\left[m_{\text{cal}}c_{\text{Cu}} + m_{\text{water}}c_{\text{water}}\right]\left(0°\text{C} - T_{i,\text{hot}}\right),$$

or

$$Q_{\text{available}} = \left[(0.080\ \text{kg})(387\ \text{J/kg}\cdot°\text{C}) + (0.560\ \text{kg})(4\,186\ \text{J/kg}\cdot°\text{C})\right](25°\text{C}) = 5.9\times10^{4}\ \text{J}$$

and we see that the warmer materials can provide more than the minimum energy needed to melt the ice. The final temperature will be between 0°C and 25°C, so we do a standard calorimetry calculation to determine T_f, recognizing that the specific heat of the ice will be the same as that of liquid water after the ice has melted. With the system thermally isolated from the environment, $Q_{\text{cold}} = -Q_{\text{hot}}$, or

$$m_{\text{ice}}\left\{c_{\text{ice}}\left[0°\text{C} - (-78°\text{C})\right] + L_f + c_{\text{water}}\left(T_f - 0°\text{C}\right)\right\} = -\left(m_{\text{water}}c_{\text{water}} + m_{\text{cal}}c_{\text{Cu}}\right)\left(T_f - 25°\text{C}\right)$$

giving

$$T_f = \frac{\left(m_{\text{water}}c_{\text{water}} + m_{\text{cal}}c_{\text{Cu}}\right)(25°\text{C}) - m_{\text{ice}}\left[c_{\text{ice}}(78°\text{C}) + L_f\right]}{\left(m_{\text{water}} + m_{\text{ice}}\right)c_{\text{water}} + m_{\text{cal}}c_{\text{Cu}}}$$

With

$$m_{\text{ice}} = 0.040\ \text{kg},\quad m_{\text{cal}} = 0.080\ \text{kg},\quad m_{\text{water}} = 0.560\ \text{kg},\quad L_f = 3.33\times10^{5}\ \text{J/kg}$$

$$c_{\text{ice}} = 2\,090\ \text{J/kg°C},\quad c_{\text{Cu}} = 387\ \text{J/kg°C},\quad \text{and}\quad c_{\text{water}} = 4\,186\ \text{J/kg°C, we have}$$

$$T_f = \frac{\left[(0.560)(4\,186) + (0.080)(387)\right](25°\text{C}) - (0.040)\left[(2\,090)(78°\text{C}) + 3.33\times10^{5}\right]}{(0.560 + 0.040)(4\,186) + 0.080(387)}$$

which yields

$$T_f = 16°\text{C}$$ ◊

---

33. A high-end gas stove usually has at least one burner rated at 14 000 Btu/h. (a) If you place a 0.25-kg aluminum pot containing 2.0 liters of water at 20°C on this burner, how long will it take to bring the water to a boil, assuming all the heat from the burner goes into the pot? (b) Once boiling begins, how much time is required to boil all the water out of the pot?

Solution

(a) The mass of 2.0 liters of water is

$$m = \rho V = \left(1.00\times10^{3}\ \frac{\text{kg}}{\cancel{\text{m}^3}}\right)\left(2.0\ \cancel{\text{L}}\right)\left(\frac{1.00\times10^{-3}\ \cancel{\text{m}^3}}{1\ \cancel{\text{L}}}\right) = 2.0\ \text{kg}$$

The thermal energy required to raise the temperature of the water and its aluminum container from 20°C to the boiling point of water at 100°C (assuming a pressure of 1 atm) is $Q_{\text{boil}} = \left(m_{\text{water}}c_{\text{water}} + m_{\text{Al}}c_{\text{Al}}\right)\Delta T$. Obtaining the needed specific heats from Table 11.1 in the text, we find

$$Q_{\text{boil}} = \left[(2.0\ \text{kg})(4\,186\ \text{J/kg}\cdot{}^{\circ}\text{C}) + (0.25\ \text{kg})(900\ \text{J/kg}\cdot{}^{\circ}\text{C})\right](100^{\circ}\text{C} - 20^{\circ}\text{C})$$

$$= 6.9\times10^{5}\ \text{J}$$

The time required for the 14 000 Btu/h burner to deliver this quantity of energy is

$$t = \frac{Q_{\text{boil}}}{P} = \frac{6.9\times10^{5}\ \text{J}}{1.4\times10^{4}\ \text{Btu/h}}\left(\frac{1\ \text{Btu/h}}{0.293\ \text{J/s}}\right) = 1.7\times10^{2}\ \text{s}\left(\frac{1\ \text{min}}{60\ \text{s}}\right) = 2.8\ \text{min} \quad \Diamond$$

(b) Once the water and its container have reached the boiling temperature of the water, the quantity of thermal energy that must be added to evaporate all of the water (with no change in temperature) is $Q_{\text{evaporate}} = m_{\text{water}}L_{v,\text{water}}$. Using Table 11.2 in the text to find the heat of vaporization of water, we have

$$Q_{\text{evaporate}} = (2.0\ \text{kg})\left(2.26\times10^{6}\ \text{J/kg}\right) = 4.5\times10^{6}\ \text{J}$$

The time the burner needs to deliver this amount of thermal energy is

$$t = \frac{Q_{\text{evaporate}}}{P} = \frac{4.5\times10^{6}\ \text{J}}{1.4\times10^{4}\ \text{Btu/h}}\left(\frac{1\ \text{Btu/h}}{0.293\ \text{J/s}}\right) = 1.1\times10^{3}\ \text{s}\left(\frac{1\ \text{min}}{60\ \text{s}}\right) = 18\ \text{min} \quad \Diamond$$

44. A thermopane window consists of two glass panes, each 0.50 cm thick, with a 1.0-cm-thick sealed layer of air in between. (a) If the inside temperature is 23°C and the outside surface temperature is 0.0°C, determine the rate of energy transfer through 1.0 m^2 of the window. (b) Compare your answer to (a) with the rate of energy transfer through 1.0 m^2 of a single 1.0-cm-thick pane of glass.

Solution

(a) The rate of energy transfer through a slab made of multiple layers of materials may be written as $Q/\Delta t = A(T_h - T_c)/R$, where A is the surface area of one side of the slab, and $(T_h - T_c)$ is the total change in temperature going from one side of the slab to the other. The R value for the slab is $R = \Sigma L_i/\kappa_i$, where L_i is the thickness of the ith layer in the compound slab and κ_i is the thermal conductivity of the material making up that layer. The R value for the thermopane window is

$$R = \frac{L_{\text{pane}}}{\kappa_{\text{glass}}} + \frac{L_{\text{trapped air}}}{\kappa_{\text{air}}} + \frac{L_{\text{pane}}}{\kappa_{\text{glass}}} = 2\left(\frac{L_{\text{pane}}}{\kappa_{\text{glass}}}\right) + \frac{L_{\text{trapped air}}}{\kappa_{\text{air}}}$$

Using Table 11.3 in the text to obtain the thermal conductivities of glass and air, we find

$$R = 2\left(\frac{0.50\times10^{-2}\ \text{m}}{0.8\ \text{W/m}\cdot{}^\circ\text{C}}\right) + \frac{1.0\times10^{-2}\ \text{m}}{0.023\,4\ \text{W/m}\cdot{}^\circ\text{C}} = (0.01+0.43)\ \frac{\text{m}^2\cdot{}^\circ\text{C}}{\text{W}} = 0.44\ \frac{\text{m}^2\cdot{}^\circ\text{C}}{\text{W}}$$

Thus, with a temperature change of $(T_h - T_c) = 23°\text{C} - 0°\text{C}$ across it and a surface area $A = 1.0\ \text{m}^2$, the rate of energy transfer through the thermopane is

$$P_{\text{thermopane}} = \left(\frac{Q}{\Delta t}\right) = \frac{(1.0\ \text{m}^2)(23°\text{C} - 0°\text{C})}{0.44\ \text{m}^2\cdot{}^\circ\text{C/W}} = 52\ \text{W} \quad \Diamond$$

(b) For the single-pane window, the R value of the thick glass pane is

$$R_{\text{thick pane}} = \frac{1.0\times10^{-2}\ \text{m}}{0.8\ \text{m}^2\cdot{}^\circ\text{C/W}} = 1\times10^{-2}\ \text{m}\cdot{}^\circ\text{C/W}$$

and the rate of energy transfer through this window is

$$P_{\text{thermopane}} = \left(\frac{Q}{\Delta t}\right) = \frac{(1.0\ \text{m}^2)(23°\text{C} - 0°\text{C})}{1\times10^{-2}\ \text{m}^2\cdot{}^\circ\text{C/W}} = 2\times10^3\ \text{W} = 2\ \text{kW} \quad \Diamond$$

48. A solar sail is made of aluminized Mylar having an emissivity of 0.03 and reflecting 97% of the light that falls on it. Suppose a sail with area $1.00\ \text{km}^2$ is oriented so that sunlight falls perpendicular to its surface with an intensity of $1.40\times10^3\ \text{W/m}^2$. To what temperature will it warm before it emits as much energy (from both sides) by radiation as it absorbs on the sunny side? Assume the sail is so thin that the temperature is uniform and no energy is emitted from the edges. Take the environment temperature to be 0 K.

Solution

The rate at which solar energy arrives at the sail is given by $P_{\text{incident}} = I\cdot A$, where I is the intensity of the solar radiation and A is the area exposed to the sunlight. This gives

$$P_{\text{incident}} = \left(1.40\times10^3\ \text{W/m}^2\right)\left(1.00\ \text{km}^2\right)\left(\frac{10^3\ \text{m}}{1\ \text{km}}\right)^2 = 1.40\times10^9\ \text{W}$$

Of this incident radiation, 97% is reflected, and only 3.0% is absorbed. The rate at which the sail absorbs solar energy is then

$$P_{\text{absorbed}} = 0.030\cdot P_{\text{incident}} = 0.030\left(1.40\times10^9\ \text{W}\right) = 4.2\times10^7\ \text{W}$$

Assuming the sail radiates equally from both sides, the total radiating area is $A' = 2A = 2(1.00\ \text{km})^2 = 2\left(1.00\times10^3\ \text{m}\right)^2 = 2.00\times10^6\ \text{m}^2$. Stefan's law gives the rate at which this sail will radiate energy to a 0 K environment, when the sail is at absolute temperature T, as

$$P_{\text{radiation}} = \sigma A' e\left(T^4 - 0\right) = \left(5.669\,6\times10^{-8}\ \text{W/m}^2\cdot\text{K}^4\right)\left(2.00\times10^6\ \text{m}^2\right)(0.03)T^4$$

or

$$P_{\text{radiation}} = \left(3.4\times10^{-3}\ \text{W/K}^4\right)T^4$$

When the sail reaches its equilibrium temperature, it radiates energy away at the same rate as it is absorbing energy, or $P_{\text{radiation}} = P_{\text{absorbed}}$. The equilibrium temperature is then

$$T = \left[\frac{P_{\text{absorbed}}}{3.4\times10^{-3}\ \text{W/K}^4}\right]^{1/4} = \left[\frac{4.2\times10^7\ \text{W}}{3.4\times10^{-3}\ \text{W/K}^4}\right]^{1/4} = 330\ \text{K} \qquad \Diamond$$

55. A 200-g block of copper at a temperature of 90°C is dropped into 400 g of water at 27°C. The water is contained in a 300-g glass container. What is the final temperature of the mixture?

Solution

Assuming that the mixture is thermally isolated from the environment, the total energy absorbed by the originally cooler parts of the mixture (water and glass) must equal the energy given up by the originally hotter material (copper). Hence, $Q_{\text{water}} + Q_{\text{glass}} = -Q_{\text{Cu}}$, or

$$\left(m_{\text{water}}c_{\text{water}} + m_{\text{glass}}c_{\text{glass}}\right)\left(T_f - T_{i,\text{cold}}\right) = -m_{\text{Cu}}c_{\text{Cu}}\left(T_f - T_{i,\text{Cu}}\right)$$

which yields

$$\left(m_{\text{water}}c_{\text{water}} + m_{\text{glass}}c_{\text{glass}} + m_{\text{Cu}}c_{\text{Cu}}\right)T_f = m_{\text{Cu}}c_{\text{Cu}}T_{i,\text{Cu}} + \left(m_{\text{water}}c_{\text{water}} + m_{\text{glass}}c_{\text{glass}}\right)T_{i,\text{cold}}$$

and

$$T_f = \frac{m_{\text{Cu}}c_{\text{Cu}}T_{i,\text{Cu}} + \left(m_{\text{water}}c_{\text{water}} + m_{\text{glass}}c_{\text{glass}}\right)T_{i,\text{cold}}}{m_{\text{water}}c_{\text{water}} + m_{\text{glass}}c_{\text{glass}} + m_{\text{Cu}}c_{\text{Cu}}}$$

With

$T_{i,\text{Cu}} = 90°\text{C}$, $T_{i,\text{cold}} = 27°\text{C}$, $m_{\text{Cu}} = 0.200$ kg, $m_{\text{water}} = 0.400$ kg, $m_{\text{glass}} = 0.300$ kg,

$c_{\text{Cu}} = 387$ J/kg·°C, $c_{\text{glass}} = 837$ J/kg·°C, and $c_{\text{water}} = 4\,186$ J/kg·°C, this gives

$$T_f = \frac{(0.200)(387)(90) + \left[(0.400)(4\,186) + (0.300)(837)\right](27)}{(0.400)(4\,186) + (0.300)(837) + (0.200)(387)} = 29°\text{C}$$ ◊

---

61. A bar of gold (Au) is in thermal contact with a bar of silver (Ag) of the same length and area (Fig. P11.61). One end of the compound bar is maintained at 80.0°C, and the opposite end is at 30.0°C. Find the temperature at the junction when the energy flow reaches a steady state.

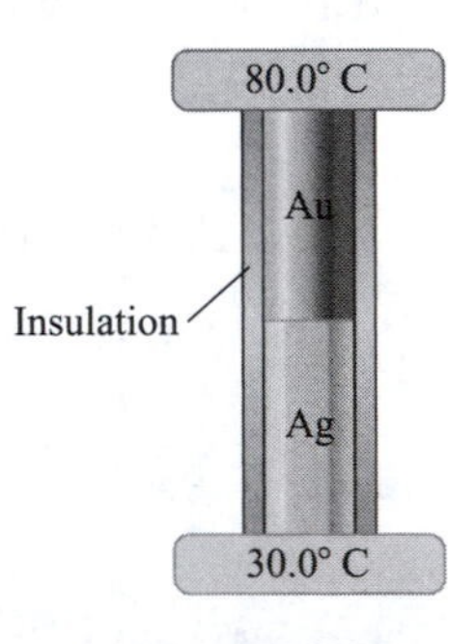

Figure P11.61

Solution

In the situation shown in Figure P11.61, thermal energy will be transported from the higher temperature (80.0°C) reservoir to the lower temperature (30.0°C) by conduction through the joined gold and silver bars. When a steady state in the energy flow has been established, energy is conducted through the gold bar to the junction between bars at the same rate it is conducted away from the junction through the silver bar. That is, at steady state, $P_{\text{Au}} = P_{\text{Ag}}$, where P_{Au} is the power transfer through the gold bar and P_{Ag} is that through the silver bar.

The rate of energy transfer by conduction through a material is given by $P = \kappa A\left(T_h - T_c\right)/L$ (see Equation 11.7 in the text), where κ is the thermal conductivity of the material, A is the cross-sectional area of the material, L is the thickness (or length if a rod or bar) of the material, and $(T_h - T_c)$ is the change in temperature maintained across the material. Thus, we have $P_{Au} = \kappa_{Au} A(80.0°C - T_{junction})/L$, and $P_{Ag} = \kappa_{Ag} A(T_{junction} - 30.0°C)/L$. Equating these and canceling the common factors A and L, we solve for the steady-state temperature at the junction to find

$$T_{junction} = \frac{\kappa_{Au}(80.0°C) + \kappa_{Ag}(30.0°C)}{\kappa_{Ag} + \kappa_{Au}}$$

The thermal conductivities for gold and silver are found in Table 11.3 of the text, giving

$$T_{junction} = \frac{(314 \text{ W/m}\cdot°C)(80.0°C) + (427 \text{ W/m}\cdot°C)(30.0°C)}{427 \text{ W/m}\cdot°C + 314 \text{ W/m}\cdot°C} = 51.2°C \qquad \lozenge$$

---

64. Three liquids are at temperatures of 10°C, 20°C, and 30°C, respectively. Equal masses of the first two liquids are mixed, and the equilibrium temperature is 17°C. Equal masses of the second and third are then mixed, and the equilibrium temperature is 28°C. Find the equilibrium temperature when equal masses of the first and third are mixed.

Solution

We are given that the masses of the liquid samples are $m_1 = m_2 = m_3$, and that the initial temperatures of the samples are $T_{i,1} = 10°C$, $T_{i,2} = 20°C$, and $T_{i,3} = 30°C$.

We assume that all of the mixtures are isolated from the environment, so $Q_{cold} = -Q_{hot}$ in each case.

When liquid 1 and liquid 2 are mixed, $T_f = 17°C$, giving

$$\cancel{m_1} c_1(17°C - 10°C) = -\cancel{m_2} c_2(17°C - 20°C) \qquad \text{or} \qquad c_1 = \frac{3}{7}c_2 \qquad [1]$$

When liquid 2 and liquid 3 are mixed, $T_f = 28°C$, and we have

$$\cancel{m_2} c_2(28°C - 20°C) = -\cancel{m_3} c_3(28°C - 30°C) \qquad \text{or} \qquad c_3 = \frac{8}{2}c_2 = 4c_2 \qquad [2]$$

Then, when liquids 1 and 3 are mixed, we would find

$$\cancel{m_1} c_1(T_f - 10°C) = -\cancel{m_3} c_3(T_f - 30°C)$$

or

$$T_f = \frac{c_1(10°C) + c_3(30°C)}{c_1 + c_3}$$

Substituting from equations [1] and [2] yields

$$T_f = \frac{(3/7)\cancel{c_2}(10°C) + (4)\cancel{c_2}(30°C)}{(3/7)\cancel{c_2} + (4)\cancel{c_2}} = \frac{3(10°C) + 28(30°C)}{3 + 28} = 28°C \qquad \lozenge$$

---

71. At time $t = 0$, a vessel contains a mixture of 10 kg of water and an unknown mass of ice in equilibrium at 0°C. The temperature of the mixture is measured over a period of an hour, with the following results: During the first 50 min, the mixture remains at 0°C; from 50 min to 60 min, the temperature increases steadily from 0°C to 2°C. Neglecting the heat capacity of the vessel, determine the mass of ice that was initially placed in it. Assume a constant power input to the container.

Solution

When thermal energy is added to a mixture of ice and water, all at the freezing point of water, the added energy initially goes into melting ice, and the temperature of the mixture will not begin to rise until all of the ice has been converted to liquid water. Once the ice is melted, the temperature of the melted ice and the original water starts to rise as energy continues to be added.

Let m be the mass of ice in the original mixture and P be the constant rate of adding thermal energy (or the power input). The thermal energy that must be added to melt all of the ice is $Q_1 = mL_f = m(3.33 \times 10^5 \text{ J/kg})$, where L_f is the latent heat of fusion of water. This quantity of energy equals the entire energy input during the first 50 minutes, or

$$Q_1 = m(3.33 \times 10^5 \text{ J/kg}) = P(50 \text{ min})$$

so the constant power input may be expressed as

$$P = \frac{Q_1}{50 \text{ min}} = \frac{m(3.33 \times 10^5 \text{ J/kg})}{50 \text{ min}} \qquad [1]$$

The energy input in the last 10 minutes produces an increase in temperature of $\Delta T = 2.0°\text{C}$ in the $m_{\text{total}} = m + 10 \text{ kg}$ of liquid water, so we may write

$$Q_2 = m_{\text{total}} c_{\text{water}} (\Delta T) = (m + 10 \text{ kg})(4\,186 \text{ J/kg} \cdot °\text{C})(2.0°\text{C}) = P(10 \text{ min})$$

or

$$P = \frac{(m + 10 \text{ kg})(4186 \text{ J/kg} \cdot °\text{C})(2.0°\text{C})}{10 \text{ min}} \qquad [2]$$

Equating the two expressions for the constant power (Equations [1] and [2]) and simplifying gives

$$m + 10 \text{ kg} = \frac{m(3.33 \times 10^5 \text{ J/kg})(10 \text{ min})}{(50 \text{ min})(4186 \text{ J/kg} \cdot °\text{C})(2.0°\text{C})} = 8.0m$$

This reduces to $7.0m = 10$ kg and yields $m = 10 \text{ kg}/7.0 = 1.4$ kg. ◊

12

The Laws of Thermodynamics

NOTES FROM SELECTED CHAPTER SECTIONS

12.1 Work in Thermodynamic Processes

The work done on a gas in a process that takes it from some initial state to some final state is the negative of the area under the curve on a *PV* diagram. (See the figure to the right.)

If the gas is compressed ($V_f < V_i$), the work done on the gas is positive.

If the gas expands ($V_f > V_i$), the work done on the gas is negative.

If the gas expands at constant pressure**,** called an **isobaric process**, then $W = -P(V_f - V_i)$.

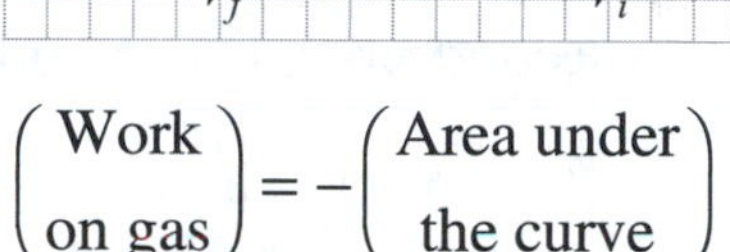

The work done on a system depends on the process (path) by which the system goes from the initial to the final state; the quantity of work depends on the initial, final, and intermediate states of the system.

12.2 The First Law of Thermodynamics
12.3 Thermal Processes

In the **first law of thermodynamics**, $\Delta U = Q + W$, the quantity Q is the energy transferred to the system by heat and W is the work done on the system. Note that by convention, Q is positive when energy enters the system and negative when energy is removed from the system. Likewise, W can be positive or negative as explained in the paragraph above. The initial and final states must be equilibrium states; however, the intermediate states are, in general, non-equilibrium states since the thermodynamic coordinates undergo finite changes during the thermodynamic process.

Four basic thermodynamic processes as they apply to an ideal gas will be described in this chapter. The *PV* diagrams associated with the processes are identified by labels (A, B, C, D).

A = isochoric (constant volume)

B = adiabatic (no thermal energy transfer)

C = isothermal (constant temperature)

D = isobaric (constant pressure)

12.4 Heat Engines and the Second Law of Thermodynamics

A **heat engine** is a device that converts internal energy to other useful forms, such as electrical and mechanical energy. A heat engine carries some working substance through a cyclic process during which (1) energy is transferred from a reservoir at a high temperature, (2) work is done by the engine, and (3) energy is expelled by the engine to a reservoir at a lower temperature.

The engine absorbs a quantity of energy, Q_h, from a hot reservoir, does work W_{eng}, and then gives up energy Q_c to a cold reservoir. Because the working substance goes through a cycle, its initial and final internal energies are equal, so $\Delta U = 0$. Hence, from the first law of thermodynamics, ***the work done by a heat engine equals the net energy absorbed from the reservoirs:*** $W_{\text{eng}} = Q_h - Q_c$.

If the working substance is a gas, the ***work done by the engine*** *for a cyclic process is the area enclosed by the curve representing the process on a PV diagram.*

The **thermal efficiency**, *e*, of a heat engine is the ratio of the work done by the engine to the energy absorbed at the higher temperature during one cycle.

The second law of thermodynamics can be stated as follows:

> It is impossible to construct a heat engine that, operating in a cycle, produces no other effect than the absorption of heat from a reservoir and the performance of an equal amount of work.

A process is **irreversible** if the system and its surroundings cannot be returned to their initial states. A process is **reversible** if the system passes from the initial to the final state through a succession of equilibrium states and can be returned to its initial condition along the same path.

The **Carnot cycle** is the most efficient cyclic process (or engine) operating between two given energy reservoirs with temperatures T_h and T_c. The *PV* diagram for the Carnot cycle, illustrated in the figure, includes **four reversible processes: two isothermal and two adiabatic.**

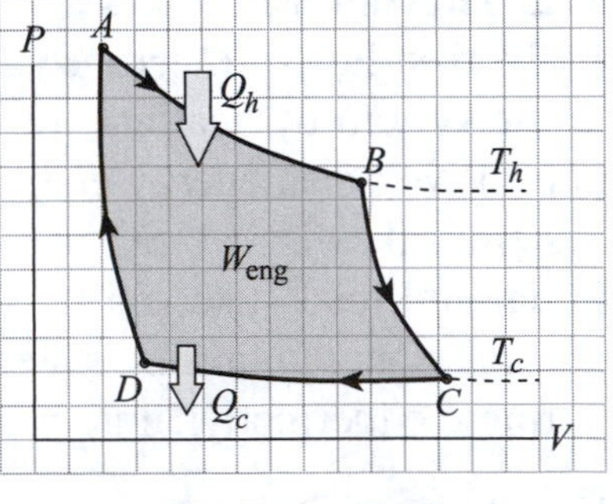

The Carnot Cycle

1. The process $A \rightarrow B$ is an isotherm (constant T). During this process the gas expands at constant temperature T_h and absorbs energy Q_h from the hot reservoir.

2. The process $B \rightarrow C$ is an adiabatic expansion ($Q = 0$); the gas expands and cools to a temperature T_c.

3. The process $C \rightarrow D$ is a second isotherm (constant T); the gas is compressed at constant temperature T_c, and expels energy Q_c to the cold reservoir.

4. The final process $D \rightarrow A$ is an adiabatic compression ($Q = 0$) in which the gas temperature increases to a final temperature of T_h.

In practice, no working engine is 100% efficient, even when losses such as friction are neglected. One can obtain some theoretical limits on the efficiency of a real engine by comparison with the ideal Carnot engine. A **reversible engine** is one which will operate with the same efficiency in the forward and reverse directions. The Carnot engine is one example of a reversible engine.

> All Carnot engines operating reversibly between T_h and T_c have the same efficiency given by Equation 12.16.
>
> No real (irreversible) engine can have an efficiency greater than that of a reversible engine operating between the same two temperatures.

12.5 Entropy

Entropy is a quantity used to measure the **degree of disorder** in a system. For example, the molecules of a gas in a container at a high temperature are in a more disordered state (higher entropy) than the same molecules at a lower temperature.

When energy is transferred to a system by heat, the entropy increases. When energy is transferred out of a system by heat, the entropy decreases. In describing a thermodynamic process, the change in entropy is an important quantity; the entropy concept is quite powerful when analyzing a system that undergoes a change in state.

The second law of thermodynamics can be stated in terms of entropy as follows:

> *The total entropy of an isolated system always increases in time if the system undergoes an irreversible process. If an isolated system undergoes a reversible process, the total entropy remains constant.*

EQUATIONS AND CONCEPTS

Work done on a gas as the volume changes is equal to the negative of the area under the pressure vs. volume curve. When ΔV is negative (compression), the work done on the gas is positive; when ΔV is positive (expansion), the work done on the gas is negative. *Equation 12.1 applies if the pressure of the gas remains constant during a compression or an expansion.*

$$W = -P\Delta V \tag{12.1}$$

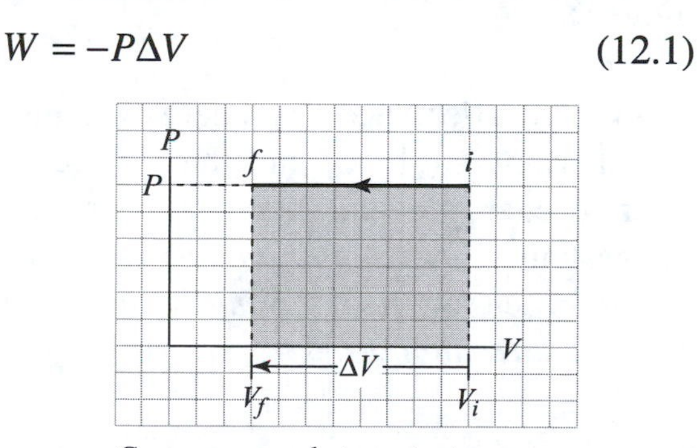

Gas compressed at constant pressure.

The **first law of thermodynamics** is stated in mathematical form in Equation 12.2. The change in the internal energy of a system is equal to the sum of the energy transferred across the boundary by heat and the energy transferred by work. Q is positive when energy is added to the system by heat. W is positive when work is done on the system by its surroundings. *The values of both Q and W depend on the path or sequence of processes by which a system changes from an initial to a final state. The internal energy U depends only on the initial and final states of the system.*

$$\Delta U = U_f - U_i = Q + W \quad (12.2)$$

Q = energy added to system by heat

W = work done on system

ΔU = change in the internal energy of system

Applications of the laws of thermodynamics will be described using some of the terms and equations shown below.

An **isolated system** does not interact with the surroundings. *The internal energy of an isolated system remains constant.*

$$Q = W = 0$$

$$\Delta U = 0$$

In a **cyclic process** the initial and final states are the same and there is no change in the internal energy of the system. *The net work done* ***on*** *the gas per cycle equals the* ***negative*** *of the area enclosed by the path representing the process on a PV diagram.*

$$\Delta U = 0$$

$$Q = -W$$

In an **adiabatic process** no energy is transferred by heat ($Q = 0$) and the change in internal energy equals the work done on the system. *A system can undergo an adiabatic process if it is thermally insulated from its surroundings.*

$$\Delta U = W$$

$$PV^\gamma = \text{constant} \quad (12.8a)$$

$$\gamma = \frac{C_p}{C_v} \quad (12.8b)$$

C_p = molar heat capacity at constant pressure

C_v = molar heat capacity at constant volume

An **isobaric process** occurs at constant pressure, and energy (Q) is transferred into (out of) the gas when it expands (contracts).

$$W = -P(V_f - V_i)$$

$$Q = nC_p\Delta T \quad (12.6)$$

An **isovolumetric process** occurs at constant volume ($\Delta V = 0$), and zero work is done. *The net energy added to the system by heat at constant volume goes into increasing the internal energy.*

$$\Delta U = Q$$

$$Q = nC_v \Delta T \qquad (12.9)$$

An **isothermal process** occurs at constant temperature. *There is no change in internal energy ($\Delta U = 0$), and the work done on the system is equal to the negative of the energy added by heat.*

$$W = -Q$$

$$W_{\text{env}} = nRT \ln\left(\frac{V_f}{V_i}\right) \qquad (12.10)$$

W_{env} = work done on the environment

A **heat engine** is a device that converts thermal energy into other forms of energy by carrying a working substance through a cyclic process. During each cycle, the following occurs:

(1) Energy is transferred by heat from a high temperature reservoir.

(2) Work is accomplished by the engine.

(3) Energy is expelled to a low temperature reservoir.

The net work (W_{eng}) done by a heat engine equals the net energy absorbed by the engine.

$$W_{\text{eng}} = |Q_h| - |Q_c| \qquad (12.11)$$

Q_h is the quantity of energy absorbed from the high temperature reservoir.

Q_c is the quantity of energy expelled to the low temperature reservoir.

The **thermal efficiency** of a heat engine is the ratio of the work done to the energy absorbed at the higher temperature during one cycle of the process.

$$e \equiv \frac{W_{\text{eng}}}{|Q_h|} = \frac{|Q_h| - |Q_c|}{|Q_h|} = 1 - \frac{|Q_c|}{|Q_h|} \qquad (12.12)$$

Carnot efficiency is the maximum theoretical efficiency of a heat engine operating in an ideal reversible cycle between two reservoirs of temperatures T_h and T_c. *All real engines are less efficient than the Carnot engine.*

$$e_C = 1 - \frac{T_c}{T_h} \qquad (12.16)$$

The **coefficient of performance (COP)** is a measure of the effectiveness of a refrigerator or heat pump.

$$\text{COP(cooling mode)} = \frac{|Q_c|}{W} \qquad (12.13)$$

$$\text{COP(heating mode)} = \frac{|Q_h|}{W} \qquad (12.14)$$

Entropy, *S*, is a thermodynamic variable which characterizes the degree of disorder in a system. *All physical processes tend toward a state of increasing entropy.*

The **change in entropy** as a system goes from an initial to a final state is the ratio of the energy transferred to the system along a reversible path to the absolute temperature of the system.

$$\Delta S \equiv \frac{\Delta Q_r}{T} \tag{12.17}$$

Important formulas related to the basic thermodynamic processes are summarized below for convenient reference.

Process	**ΔU**	**Q**	**W**
Isobaric	$nC_v \Delta T$	$nC_p \Delta T$	$-P\Delta V$
Adiabatic	$nC_v \Delta T$	0	ΔU
Isovolumetric	$nC_v \Delta T$	ΔU	0
Isothermal	0	$-W$	$-nRT \ln\left(\frac{V_f}{V_i}\right)$

REVIEW CHECKLIST

- Understand how work is defined when a system undergoes a change in state, and the fact that work depends on the path taken by the system. You should also know how to sketch processes on a *PV* diagram, and calculate work using these diagrams.
- State the first law of thermodynamics ($\Delta U = Q + W$), and explain the meaning of the three forms of energy contained in this statement. Discuss the implications of the first law of thermodynamics as applied to (i) an isolated system, (ii) a cyclic process, (iii) an adiabatic process, and (iv) an isothermal process.
- Describe the processes via which an ideal heat engine goes through a **Carnot cycle**. Express the efficiency of an ideal heat engine (Carnot engine) as a function of work and energy exchange with its environment. Express the maximum efficiency of an ideal heat engine as a function of its input and output temperatures.
- Understand the concept of entropy. Define change in entropy for a system in terms of its energy gain or loss by heat and its temperature. State the second law of thermodynamics as it applies to entropy changes in a thermodynamic system.

SOLUTIONS TO SELECTED END-OF-CHAPTER PROBLEMS

7. A sample of helium behaves as an ideal gas as it is heated at constant pressure from 273 K to 373 K. If 20.0 J of work is done by the gas during this process, what is the mass of helium present?

Solution

When pressure is constant, the ideal gas law gives $PV_f - PV_i = nRT_f - nRT_i$, or $P(\Delta V) = nR(\Delta T)$. But, in an isobaric process, the work a system does on its environment is $W_{\text{env}} = P(\Delta V)$, so

$$W_{\text{env}} = nR(\Delta T) \qquad \textbf{[1]}$$

where $n = m/M$ is the number of moles present, m is the mass of the gas, and M is the molecular weight. For helium, $M_{\text{He}} = 4.00$ g/mol, and if the work done by the gas is $W_{\text{env}} = 20.0$ J, we have

$$W_{\text{env}} = \left(\frac{m}{M_{\text{He}}}\right)R(\Delta T)$$

or

$$m = \frac{W_{\text{env}} M_{\text{He}}}{R(\Delta T)}$$

which yields

$$m = \frac{(20.0\text{ J})(4.00\text{ g/mol})}{(8.31\text{ J/mol}\cdot\text{K})(373\text{ K} - 273\text{ K})} = 9.63\times10^{-2}\text{ g} = 96.3\text{ mg} \qquad \Diamond$$

As an alternate approach to this problem, we could make use of the properties of an ideal gas and the first law of thermodynamics. For an ideal gas, the energy added by heat is $Q = nC_p(\Delta T)$, and the change in the internal energy is $\Delta U = nC_v(\Delta T)$. Here, C_v is the molar heat capacity at constant volume, and $C_p = C_v + R$ is the molar heat capacity at constant pressure for the ideal gas. For a monatomic ideal gas, such as helium, $C_v = \frac{3}{2}R$ and $C_p = \frac{5}{2}R$.

The first law of thermodynamics is $\Delta U = Q + W = Q - W_{\text{env}}$, where $W = -W_{\text{env}}$ is the work done *on* the system. Thus, for this monatomic, ideal gas, we have

$$W_{\text{env}} = Q - \Delta U = n\left(\tfrac{5}{2}R\right)\Delta T - n\left(\tfrac{3}{2}R\right)\Delta T$$

or

$$W_{\text{env}} = nR(\Delta T)$$

which is the same as Equation [1] above. Thus, the two methods of solution are fully equivalent and will yield the same results.

10. (a) Determine the work done *on* a fluid that expands from i to f as indicated in Figure P12.10. (b) How much work is done *on* the fluid if it is compressed from f to i along the same path?

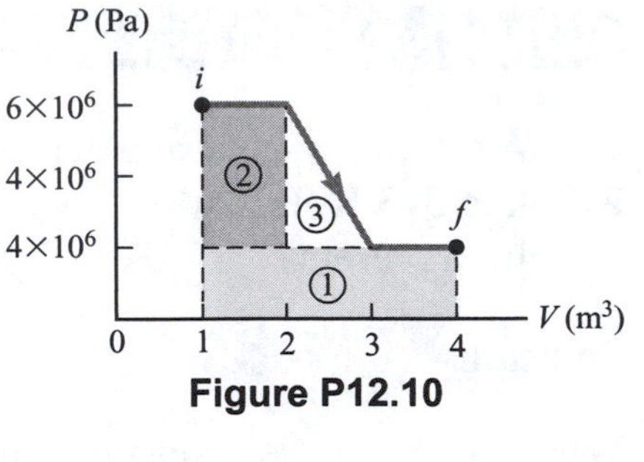

Figure P12.10

Solution

(a) When the working substance of a thermodynamic system expands (volume increases), it does work, W_{env}, on its surroundings or environment. The work done *on* the system is the negative of the work it does on the environment, $W = -W_{env}$. The magnitude of this work is given by the area under the process curve in a PV diagram as shown above. In this case, we see that this area can be broken up into three simple geometric shapes, two rectangles and one triangle. Thus, $W = -(A_1 + A_2 + A_3)$, or

$$W = -[(2\times10^6 \text{ Pa} - 0)(4 \text{ m}^3 - 1 \text{ m}^3) + (6\times10^6 \text{ Pa} - 2\times10^6 \text{ Pa})(2 \text{ m}^3 - 1 \text{ m}^3)$$

$$+\frac{1}{2}(6\times10^6 \text{ Pa} - 2\times10^6 \text{ Pa})(3 \text{ m}^3 - 2 \text{ m}^3)]$$

yielding

$$-12\times10^6 \text{ J} = -12 \text{ MJ}$$ ◊

(b) If the working substance of the system is compressed into a smaller volume, the environment does *positive* work on the system and adds energy into the system. Again, the magnitude of the work done is equal to the area under the process curve in the PV diagram. Thus, if the fluid in our system is compressed from state f to state i along the path shown in the PV diagram given above, the work done *on* the fluid is

$$W = +(A_1 + A_2 + A_3) = +12\times10^6 \text{ J} = +12 \text{ MJ}$$ ◊

---

17. A gas is enclosed in a container fitted with a piston of cross-sectional area 0.150 m^2. The pressure of the gas is maintained at 6 000 Pa as the piston moves inward 20.0 cm. (a) Calculate the work done by the gas. (b) If the internal energy of the gas decreases by 8.00 J, find the amount of energy removed from the system by heat during the compression.

Solution

(a) In an isobaric compression such as this, the work done *on* the system is $W = -P(\Delta V)$. For this piston and cylinder, the change in volume of the trapped gas is $\Delta V = -A\cdot\Delta x$, where Δx is the distance the piston moves inward. Thus,

$$W = -(6.00\times10^3 \text{ Pa})[-(0.150 \text{ m}^3)(0.200 \text{ m})] = +180 \text{ J}$$

The work done *by* the gas is then

$$W_{env} = -W = -(180 \text{ J}) = -180 \text{ J}$$ ◊

(b) The first law of thermodynamics says that the change in the internal energy of a system, ΔU, equals the sum of the energy transferred to the system by heat, Q, and the work done *on* the system, W. Thus, for the piston-cylinder-gas system of this problem, $\Delta U = Q + W$, or the energy added by heat is

$$Q = \Delta U - W = -8.00\text{ J} - (+180\text{ J}) = -188\text{ J}$$

The energy transferred from the system to the environment (i.e., removed from the system) by heat is then

$$Q_{\text{env}} = -Q = -(-188\text{ J}) = +188\text{ J} \qquad \Diamond$$

---

21. An ideal monatomic gas expands isothermally from 0.500 m^3 to 1.25 m^3 at a constant temperature of 675 K. If the initial pressure is 1.00×10^5 Pa, find (a) the work done on the gas, (b) the thermal energy transfer Q, and (c) the change in the internal energy.

Solution

(a) When an ideal gas expands isothermally, the work it does on the environment is given by Equation 12.10 in the textbook as $W_{\text{env}} = nRT\ln\left(V_f/V_i\right)$, where $T = T_i = T_f$ is the constant temperature of the gas, while V_i and V_f are the initial and final volumes, respectively. From the ideal gas law, $PV = nRT$, we may write the product nRT as $nRT = nRT_i = P_iV_i$, so the work done *on* the gas becomes $W = -W_{\text{env}} = -P_iV_i\ln\left(V_f/V_i\right)$, and gives

$$W = -\left(1.00\times10^5\text{ Pa}\right)\left(0.500\text{ m}^3\right)\ln\left(\frac{1.25\text{ m}^3}{0.500\text{ m}^3}\right) = -4.58\times10^4\text{ J} \qquad \Diamond$$

With $W < 0$, the work done on the environment $W_{\text{env}} = -W$ is positive, meaning the transfer of energy as work actually goes from the gas to the environment in this process.

(b) The first law of thermodynamics states that $\Delta U = Q + W$, where ΔU is the change in the internal energy of the system, Q is the energy transferred into the system by heat, and W is the work done on the system by the environment. When the working substance of the system is a monatomic ideal gas, the change in the internal energy is directly proportional to the change in temperature as given by Equation 12.3b in the textbook, $\Delta U = \frac{3}{2}nR\Delta T$. Since $\Delta T = 0$ in this isothermal expansion process, $\Delta U = 0$, and the first law gives the thermal energy transfer as

$$Q = -W = -\left(-4.58\times10^4\text{ J}\right) = +4.58\times10^4\text{ J} \qquad \Diamond$$

Since $Q > 0$, energy is transferred *into* the gas by heat during this process.

(c) As stated above, the change in the internal energy of a monatomic ideal gas is given by $\Delta U = \frac{3}{2}nR\Delta T$. Thus, for all isothermal processes involving a system whose working substance approximates an ideal gas, $\Delta U = 0$. $\Diamond$

---

27. Consider the Universe to be an adiabatic expansion of atomic hydrogen gas. (a) Use the ideal gas law and Equation 12.8a to show that $TV^{\gamma-1} = C$, where C is a constant. (b) The current Universe extends at least 15 billion light-years in all directions (1.4×10^{26} m), and the current temperature of the Universe is 2.7 K. Estimate the temperature of the Universe when it was the size of a nutshell, with a radius of 2 cm. (For this calculation, assume the Universe is spherical.)

Solution

(a) The relation between the pressure and the volume of an ideal gas during an adiabatic process is given by Equation 12.8a (see textbook) as $PV^{\gamma} = \text{constant}$, where $\gamma = C_p/C_v$ is the ratio of the molar specific heats of the gas. From the ideal gas law, we may write the pressure as $P = nRT/V$. Substituting this into the earlier relation gives

$$\left(\frac{nRT}{V}\right)V^{\gamma} = \text{constant}$$

or

$$TV^{\gamma-1} = \frac{\text{constant}}{nR}$$

Thus, $TV^{\gamma-1} = C$, where $C = \text{constant}/nR$ is a constant provided the number of moles of ideal gas in the system does not change. ◊

(b) Using the result of part (a) above, the initial and final states of an ideal gas undergoing an adiabatic expansion are related by $T_iV_i^{\gamma-1} = T_fV_f^{\gamma-1}$. Thus, the initial temperature is given by $T_i = T_f(V_f/V_i)^{\gamma-1}$. If we model the universe as an atomic hydrogen gas expanding uniformly in all directions (and hence, maintaining a spherical shape),

$$\left(\frac{V_f}{V_i}\right)^{\gamma-1} = \left(\frac{\frac{4}{3}\pi r_f^3}{\frac{4}{3}\pi r_i^3}\right)^{\gamma-1} = \left(\frac{r_f^3}{r_i^3}\right)^{\gamma-1} = \left(\frac{r_f}{r_i}\right)^{3\gamma-3}$$

and the initial temperature becomes $T_i = T_f(r_f/r_i)^{3\gamma-3}$. Atomic hydrogen is a gas of individual hydrogen atoms rather than hydrogen molecules. For a monatomic gas such as this, $\gamma = 5/3 = 1.67$ (see Table 12.1 in the textbook). Hence, $3\gamma - 3 = 2$, and we have

$$T_i = T_f\left(\frac{r_f}{r_i}\right)^2 = (2.7\text{ K})\left(\frac{1.4\times10^{26}\text{ m}}{2\times10^{-2}\text{ m}}\right)^2 = 1\times10^{56}\text{ K}$$ ◊

37. An engine absorbs 1.70 kJ from a hot reservoir at 277°C and expels 1.20 kJ to a cold reservoir at 27°C in each cycle. (a) What is the engine's efficiency? (b) How much work is done by the engine in each cycle? (c) What is the power output of the engine if each cycle lasts 0.300 s?

Solution

(a) The work done by a heat engine is equal to the net energy absorbed. Thus, if during a full cycle, the engine takes in energy Q_h from a high temperature thermal energy reservoir and exhausts a quantity Q_c to a lower temperature energy reservoir, the work done is $W_{\text{eng}} = Q_h - Q_c$. The efficiency of the engine is the ratio of the work done to the amount of energy input from the high temperature reservoir. Hence,

$$e = \frac{W_{\text{eng}}}{Q_h} = \frac{Q_h - Q_c}{Q_h} = 1 - \frac{Q_c}{Q_h} = 1 - \frac{1.20\text{ kJ}}{1.70\text{ kJ}} = 0.294 \quad \text{(or 29.4\%)}$$ ◊

(b) As defined above, the work the engine does each cycle is equal to the net energy absorbed per cycle. For this engine,

$$W_{\text{eng}} = Q_h - Q_c = 1.70\text{ kJ} - 1.20\text{ kJ} = 0.50\text{ kJ} = 0.50 \times 10^3\text{ J} = 500\text{ J}$$ ◊

(c) The power output from the engine is the rate at which it does work. Therefore, if this engine outputs 500 J per cycle and each cycle lasts 0.300 s, the power output is

$$P = \frac{W_{\text{eng}}}{t} = \frac{500\text{ J}}{0.300\text{ s}} = 1.67 \times 10^3\text{ J/s} = 1.67 \times 10^3\text{ W} = 1.67\text{ kW}$$ ◊

41. In one cycle, a heat engine absorbs 500 J from a high temperature reservoir and expels 300 J to a low temperature reservoir. If the efficiency of this engine is 60% of the efficiency of a Carnot engine, what is the ratio of the low temperature to the high temperature in the Carnot engine?

Solution

The efficiency of a heat engine is $e = W_{\text{eng}}/|Q_h|$, where W_{eng} is the work done by the engine, and $|Q_h|$ is the energy drawn from the hot (or higher temperature) reservoir. But $W_{\text{eng}} = |Q_h| - |Q_c|$, where $|Q_c|$ is the energy exhausted to the cold (or lower temperature) reservoir. Thus, the efficiency may be written as

$$e = \frac{W_{\text{eng}}}{|Q_h|} = \frac{|Q_h| - |Q_c|}{|Q_h|} = 1 - \frac{|Q_c|}{|Q_h|} = 1 - \frac{300 \text{ J}}{500 \text{ J}} = 0.40 \text{ (or 40\%)}$$

If the efficiency of our actual machine is 60% of the Carnot efficiency, then $e = 0.60 \cdot e_C$, and the efficiency of a Carnot engine operating between the thermal energy reservoirs used by our engine would be

$$e_C = \frac{e}{0.60} = \frac{0.40}{0.60} = \frac{2}{3}$$

Since the efficiency of a Carnot engine can be expressed in terms of the absolute temperatures of its reservoirs as

$$e_C = \frac{T_h - T_c}{T_h} = 1 - \frac{T_c}{T_h}$$

the ratio of the temperature of the cold reservoir to that of the hot reservoir must be

$$\frac{T_c}{T_h} = 1 - e_C = 1 - \frac{2}{3} = \frac{1}{3}$$ ◊

45. A Styrofoam cup holding 125 g of hot water at 1.00×10^2 °C cools to room temperature, 20.0°C. What is the change in entropy of the room? (Neglect the specific heat of the cup and any change in temperature of the room.)

Solution

As the water in the Styrofoam cup cools from an initial temperature of $T_i = 1.00 \times 10^2\,°\text{C}$ to a final temperature of $T_f = 20.0°\text{C}$, the quantity of thermal energy it must give up to its surroundings is

$$Q = m_{\text{water}} c_{\text{water}} \left|T_f - T_i\right| = (125\text{ g})\left(1.00\ \frac{\text{cal}}{\text{g}\cdot°\text{C}}\right)\left|1.00 \times 10^2\,°\text{C} - 20.0°\text{C}\right|$$

or

$$Q = (0.125\text{ kg})\left(4186\ \frac{\text{J}}{\text{kg}\cdot°\text{C}}\right)(80.0°\text{C}) = 4.19 \times 10^4\text{ J}$$

Neglecting the specific heat and thermal capacity of the Styrofoam cup, this thermal energy must be absorbed by the air and surrounding surfaces of the room.

If a thermal reservoir, at constant absolute temperature T, absorbs a quantity of energy Q_r in a reversible process, the entropy of the reservoir changes by $\Delta S = Q_r / T$.

The gradual absorption of thermal energy from the cooling water by the room is a reversible process. Also, we assume the thermal capacity of the room is so large that its temperature changes by a negligible amount as it absorbs this energy. Then, the absolute temperature of the room is constant with a value of $T = 20.0 + 273 = 293$ K, and the change in the entropy of the room is

$$\Delta S_{\text{room}} = \frac{+Q}{T} = \frac{+4.19 \times 10^4\text{ J}}{293\text{ K}} = +143\ \text{J/K} \qquad \Diamond$$

52. When an aluminum bar is temporarily connected between a hot reservoir at 725 K and a cold reservoir at 310 K, 2.50 kJ of energy is transferred by heat from the hot reservoir to the cold reservoir. In this irreversible process, calculate the change in entropy of (a) the hot reservoir, (b) the cold reservoir, and (c) the Universe, neglecting any change in entropy of the aluminum rod. (d) Mathematically, why did the result for the Universe in part (c) have to be positive?

Solution

When a quantity of thermal energy Q is added to ($Q > 0$), or withdrawn from ($Q < 0$), a thermal energy reservoir at a constant absolute temperature T, the change in the entropy of that reservoir is $\Delta S = Q/T$.

(a) When 2.50 kJ of thermal energy is transferred out of the hot reservoir at absolute temperature $T_h = 725$ K, then $Q_h = -2.50$ kJ and the change in entropy of the hot reservoir is

$$\Delta S_h = \frac{Q_h}{T_h} = \frac{-2.50 \times 10^3 \text{ J}}{725 \text{ K}} = -3.45 \text{ J/K}$$ ◊

(b) When the 2.50 kJ of thermal energy is transferred into the cold reservoir at absolute temperature $T_c = 310$ K, we have $Q_c = +2.50$ kJ, and the change in entropy for this reservoir is

$$\Delta S_c = \frac{Q_c}{T_c} = \frac{+2.50 \times 10^3 \text{ J}}{310 \text{ K}} = +8.06 \text{ J/K}$$ ◊

(c) The change in entropy of the Universe as the result of a thermodynamic process is the sum of the changes in entropy experienced by those parts of the Universe that have undergone changes as a result of the process. Thus, neglecting any change in entropy of the aluminum bar, the total change in entropy for this process is

$$\Delta S_{\text{total}} = \Delta S_h + \Delta S_c = -3.45 \text{ J/K} + 8.06 \text{ J/K} = +4.61 \text{ J/K}$$ ◊

Note that the total change in entropy is positive, just as the second law of thermodynamics says it must be for an irreversible process.

(d) The spontaneous flow of thermal energy from one system to another is always from or *out of* the hotter system and *into* the cooler system. Thus, the total change in entropy for this process is $\Delta S_{\text{total}} = -|Q|/T_h + |Q|/T_c$, and the positive term always has the smaller denominator. Hence, the positive term will always dominate the sum, making $\Delta S_{\text{total}} > 0$. ◊

59. A 1 500-kW heat engine operates at 25% efficiency. The heat energy expelled at the low temperature is absorbed by a stream of water that enters the cooling coils at 20°C. If 60 L flows across the coils per second, determine the increase in temperature of the water.

Solution

The power output from the engine is

$$P = \frac{W_{\text{eng}}}{\Delta t} = 1\,500\text{ kW} = 1\,500 \times 10^3\text{ W} = 1.5 \times 10^6\text{ J/s}$$

so the work done by this engine in a time interval of $\Delta t = 1.0$ s is $W_{\text{eng}} = 1.5 \times 10^6$ J.

The efficiency of a heat engine is defined as $e = W_{\text{eng}}/|Q_h|$. If the efficiency of this engine is $e = 0.25\ (25\%)$, the energy absorbed by heat from the high temperature (or hot) reservoir each second is

$$|Q_h| = \frac{W_{\text{eng}}}{e} = \frac{1.5 \times 10^6\text{ J}}{0.25} = 6.0 \times 10^6\text{ J}$$

and the energy exhausted by heat to the low temperature (or cold) reservoir each second is

$$|Q_c| = |Q_h| - W_{\text{eng}} = 6.0 \times 10^6\text{ J} - 1.5 \times 10^6\text{ J} = 4.5 \times 10^6\text{ J}$$

The mass of the 60 L of 20°C water that flows over the cooling coils during this 1 second time interval is

$$m = \rho_{\text{water}} V = \left(1.0 \times 10^3\ \frac{\text{kg}}{\cancel{\text{m}^3}}\right)(60\ \cancel{\text{L}})\left(\frac{10^{-3}\ \cancel{\text{m}^3}}{1\ \cancel{\text{L}}}\right) = 60\text{ kg}$$

and the increase in the temperature of the cooling water will be

$$\Delta T = \frac{|Q_c|}{mc_{\text{water}}} = \frac{4.5 \times 10^6\text{ J}}{(60\text{ kg})(4\,186\text{ J/kg}\cdot{}^\circ\text{C})} = 18^\circ\text{C}$$ ◊

---

67. A cylinder containing 10.0 moles of a monatomic ideal gas expands from A to B along the path shown in Figure P12.67. (a) Find the temperature of the gas at point A and the temperature at point B. (b) How much work is done by the gas during this expansion? (c) What is the change in internal energy of the gas? (d) Find the energy transferred to the gas by heat in this process.

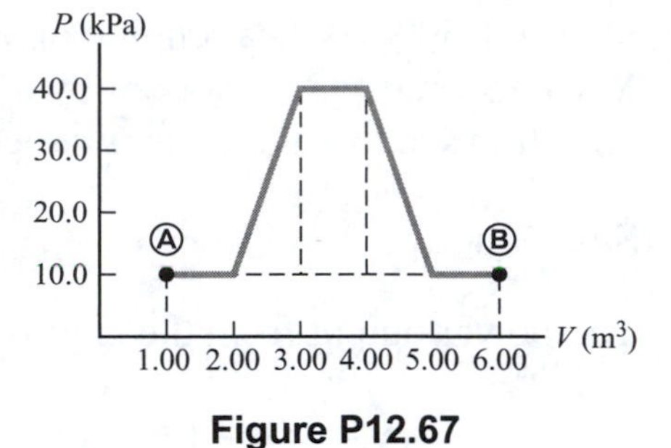

Figure P12.67

Solution

(a) At point A, the gas pressure is $P_A = 10.0 \text{ kPa} = 10.0 \times 10^3 \text{ Pa} = 1.00 \times 10^4 \text{ Pa}$ and its volume is $V_A = 1.00 \text{ m}^3$. The ideal gas law then gives its temperature as

$$T_A = \frac{P_A V_A}{nR} = \frac{(1.00 \times 10^4 \text{ Pa})(1.00 \text{ m}^3)}{10.0(8.31 \text{ J/kg} \cdot \text{K})} = 1.20 \times 10^2 \text{ K} \qquad \Diamond$$

Similarly, at point B, $P_B = 1.00 \times 10^4 \text{ Pa}$, $V_B = 6.00 \text{ m}^3$, and

$$T_B = \frac{P_B V_B}{nR} = \frac{(1.00 \times 10^4 \text{ Pa})(6.00 \text{ m}^3)}{10.0(8.31 \text{ J/kg} \cdot \text{K})} = 722 \text{ K} \qquad \Diamond$$

(b) The work done *by* the gas (since it is expanding) on the environment equals the area under the process curve between points A and B. Observe from Figure 12.67 that this area can be computed as the sum of two rectangular areas and two triangular areas. This gives

$$W_{\text{env}} = (10 \times 10^3 \text{ Pa})(6.00 \text{ m}^3 - 1.00 \text{ m}^3) + [(40.0 - 10.0) \times 10^3 \text{ Pa}](4.00 - 3.00) \text{ m}^3$$
$$+\frac{1}{2}[(40.0 - 10.0) \times 10^3 \text{ Pa}](3.00 - 2.00) \text{ m}^3$$
$$+\frac{1}{2}[(40.0 - 10.0) \times 10^3 \text{ Pa}](5.00 - 4.00) \text{ m}^3$$

or

$$W_{\text{env}} = 1.10 \times 10^5 \text{ J} \qquad \Diamond$$

(c) Since this is a monatomic, ideal gas, the change in its internal energy during this process is

$$\Delta U = nC_v \Delta T = \frac{3}{2} nR(\Delta T) = \frac{3}{2} nR(T_B - T_A)$$

or

$$\Delta U = \frac{3}{2}(10.0 \text{ mol})(8.31 \text{ J/mol} \cdot \text{K})(722 \text{ K} - 120 \text{ K}) = 7.50 \times 10^4 \text{ J} \qquad \Diamond$$

(d) Using the first law of thermodynamics, the energy transferred to the gas by heat during this process is found to be

$$Q = \Delta U - W = \Delta U - (-W_{\text{env}}) = \Delta U + W_{\text{env}}$$

or

$$Q = 7.50 \times 10^4 \text{ J} + 1.10 \times 10^5 \text{ J} = 1.85 \times 10^5 \text{ J} \qquad \Diamond$$

71. An electrical power plant has an overall efficiency of 15%. The plant is to deliver 150 MW of electrical power to a city, and its turbines use coal as fuel. The burning coal produces steam at 190°C, which drives the turbines. The steam is condensed into water at 25°C by passing through coils that are in contact with river water. (a) How many metric tons of coal does the plant consume each day $\left(1 \text{ metric ton} = 1\times 10^3 \text{ kg}\right)$? (b) What is the total cost of the fuel per year if the delivery price is $8 per metric ton? (c) If the river water is delivered at 20°C, at what minimum rate must it flow over the cooling coils in order that its temperature not exceed 25°C? (*Note:* The heat of combustion of coal is 7.8×10^3 cal/g.)

Solution

(a) If the plant has an overall efficiency of 15% and is to have a power output of 150 MW, the required power input (from combustion of coal) is

$$P_{\text{input}} = \frac{P_{\text{output}}}{e} = \frac{150\times 10^6 \text{ W}}{0.15} = 1.0\times 10^9 \text{ J/s}$$

The rate at which coal must be burned to yield this needed power input is

$$\frac{\Delta m}{\Delta t} = \frac{P_{\text{input}}}{\textit{heat of combustion}} = \frac{\left(1.0\times 10^9 \text{ J/s}\right)\left(8.64\times 10^4 \text{ s/d}\right)}{\left(7.8\times 10^3 \dfrac{\text{cal}}{\text{g}}\right)\left(\dfrac{10^3\text{g}}{1\text{kg}}\right)\left(\dfrac{4.186 \text{ J}}{1\text{cal}}\right)} = 2.6\times 10^6 \text{ kg/d}$$

or

$$\frac{\Delta m}{\Delta t} = \left(2.6\times 10^6 \frac{\text{kg}}{\text{d}}\right)\left(\frac{1 \text{ metric ton}}{10^3\text{kg}}\right) = 2.6\times 10^3 \text{ metric ton/d} \qquad \lozenge$$

(b) The annual fuel cost will be $\textit{cost} = \textit{rate}\times(\Delta m/\Delta t)\times(\Delta t)$

or

$$\textit{cost} = \left(\frac{\$8}{\text{metric ton}}\right)\left(2.6\times 10^3 \frac{\text{metric ton}}{\text{d}}\right)\left(\frac{365.242 \text{ d}}{1 \text{ y}}\right) = \$7.6\times 10^6/\text{y} \qquad \lozenge$$

(c) The rate at which cooling waters must absorb exhaust energy is

$$P_{\text{exhaust}} = P_{\text{input}} - P_{\text{output}} = 1.0\times 10^9 \text{ W} - 150\times 10^6 \text{ W} = 8.5\times 10^8 \text{ W}$$

If the maximum rise in temperature of the cooling waters is to be $\Delta T = 5.0°\text{C}$, the minimum flow rate of the water is

$$\textit{flow rate} = \frac{\Delta m_{\text{water}}}{\Delta t} = \frac{P_{\text{exhaust}}}{c_{\text{water}}(\Delta T)} = \frac{8.5\times 10^8 \text{ J/s}}{(4\,186 \text{J/kg}\cdot°\text{C})(5.0°\text{C})} = 4.1\times 10^4 \text{ kg/s} \qquad \lozenge$$

13

Vibrations and Waves

NOTES FROM SELECTED CHAPTER SECTIONS

13.1 Hooke's Law
13.2 Elastic Potential Energy

Simple harmonic motion occurs when the net force along the direction of motion is proportional to the displacement and in the opposite direction. *An object exhibits simple harmonic motion if the net external force acting on it is a linear restoring force.*

It is necessary to define a few terms relative to harmonic motion:

- **The amplitude**, A, is the maximum distance that an object moves away from its equilibrium position. In the absence of friction, an object will continue in simple harmonic motion and reach a maximum distance equal to the amplitude on each side of the equilibrium position during each cycle.

- **The period**, T, is the time it takes the object to execute one complete cycle of the motion.

- **The frequency**, f, is the number of cycles or vibrations per unit of time.

Oscillatory motions are exhibited by many physical systems such as a mass attached to a spring, a pendulum, atoms in a solid, stringed musical instruments, and electrical circuits driven by a source of alternating current. *Simple harmonic motion of a mechanical system corresponds to the oscillation of an object between two points for an indefinite period of time, with no loss in mechanical energy.*

13.4 Position, Velocity, and Acceleration as Functions of Time

The position (x), velocity (v), and acceleration (a) of an object moving with simple harmonic motion are shown in the three graphs below.

Shown here, from top to bottom, are graphs of displacement, velocity, and acceleration versus time for an object moving with simple harmonic motion under the initial conditions $x = A$ and $v = 0$ at $t = 0$.

At the instant indicated by the dashed line:

$t = (3T/4)$

$x = 0$ (moving in the +x direction)

$v = +\omega A$ (maximum positive value)

$a = 0$

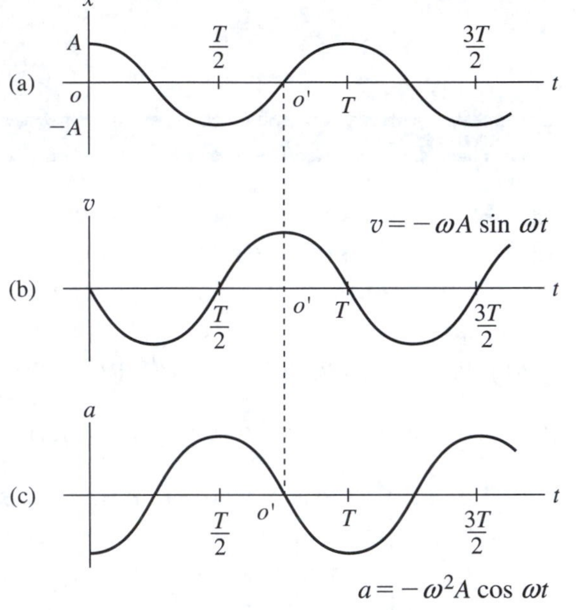

The figure on the right shows a **mass-spring system** in simple harmonic motion. The point $x = 0$ is the equilibrium position of the mass, that is, the point where the mass would reside if left undisturbed. In this position, there is no horizontal force on the mass. When the mass is displaced a distance x from its equilibrium position, the spring produces a linear restoring force given by Hooke's law, $F = -kx$. The minus sign means that $\vec{\mathbf{F}}$ is directed toward the left when the displacement, x, is positive and is directed toward the right when x is negative. *The direction of the force* $\vec{\mathbf{F}}$ *is always toward the equilibrium position.* Examine the figure and note the positions of the mass which correspond to maximum values of velocity and acceleration.

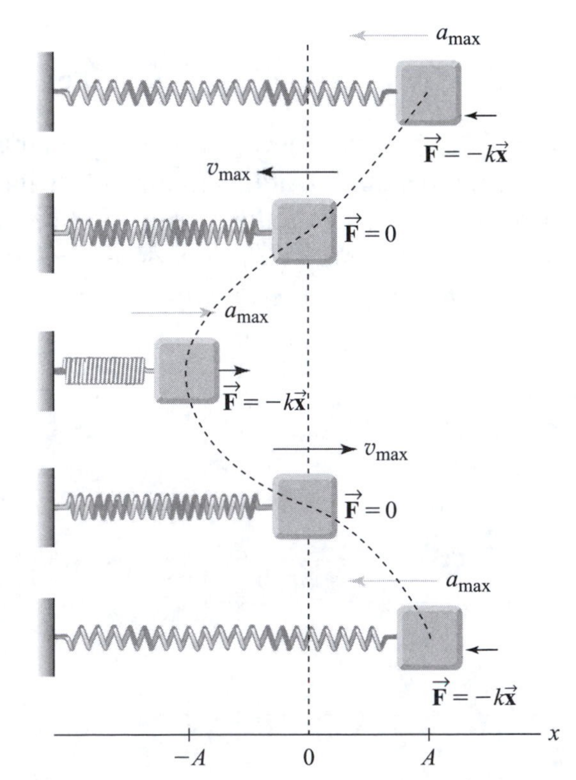

13.5 Motion of a Pendulum

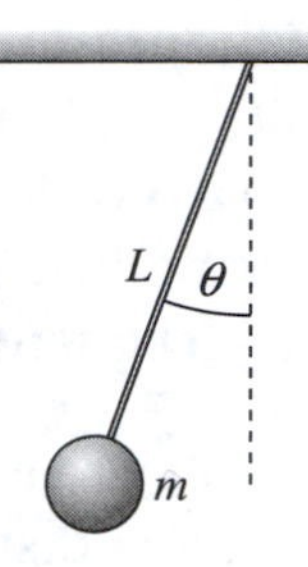

A **simple pendulum** consists of a mass m attached to a light string of length L as shown in the figure. When the maximum angular displacement is small (θ less than approximately 15°), the pendulum exhibits simple harmonic motion. In this case, the resultant force acting on the mass m equals the component of weight tangent to the circular path followed by the mass. The magnitude of the resultant force equals $mg\sin\theta$. Since this force is always directed toward $\theta = 0$, it corresponds to a restoring force.

The period depends only on the length of the pendulum and the acceleration due to gravity. The period does not depend on mass, so we conclude that all simple pendula of equal length oscillate with the same frequency or period at the same location.

13.6 Damped Oscillations

Damped oscillations occur in realistic systems in which retarding forces such as friction are present. These forces will reduce the amplitudes of the oscillations with time, since mechanical energy is continually transferred from the oscillating system. Depending on the value of the frictional (retarding) force, three distinct types of damping can be identified:

> **Underdamped**—In this case, the retarding force is small compared to the restoring force. For small damping, oscillations continue with a frequency slightly less than that of the undamped system. The amplitude of the motion decreases exponentially with time until the oscillations cease at zero amplitude.
>
> **Critically damped**—With an increase in the retarding force, the system does not oscillate and returns to the equilibrium position in the shortest possible time without passing through the equilibrium point.
>
> **Overdamped**—This mode is similar to critical damping and occurs with further increase in the frictional force. In this case, the system also returns to equilibrium without passing through the equilibrium point but requires a longer time to do so.

It is possible to compensate for the energy lost in a damped oscillator by adding an additional driving force that does positive work on the system. This additional energy supplied to the system must equal the energy lost due to friction to maintain constant amplitude.

13.7 Waves

The production of mechanical waves requires (1) an **elastic medium** that can be disturbed, (2) **an energy source** to provide a disturbance or deformation in the medium, and (3) a **physical mechanism** by way of which adjacent portions of the medium can influence each other. The three parameters important in characterizing waves are (1) **wavelength**, (2) **frequency**, and (3) **wave velocity**.

Transverse waves, shown below on the right, are those in which particles of the medium undergo displacements which are perpendicular to the direction of the wave velocity as the wave pulse moves along the rope. For **longitudinal waves**, shown on the left, the particles of the medium undergo displacements which are parallel to the direction of wave motion.

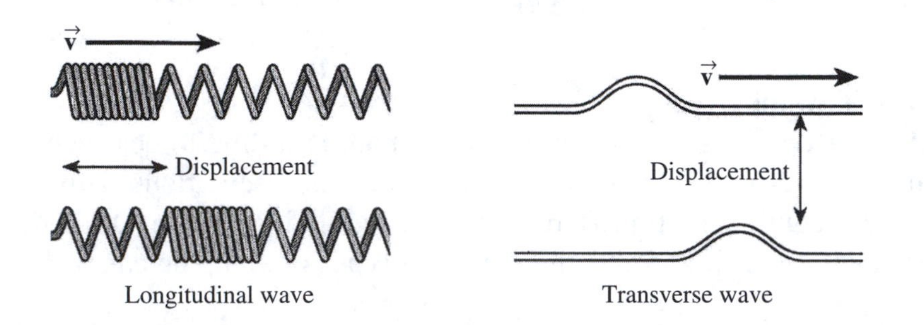

13.8 Frequency, Amplitude, and Wavelength

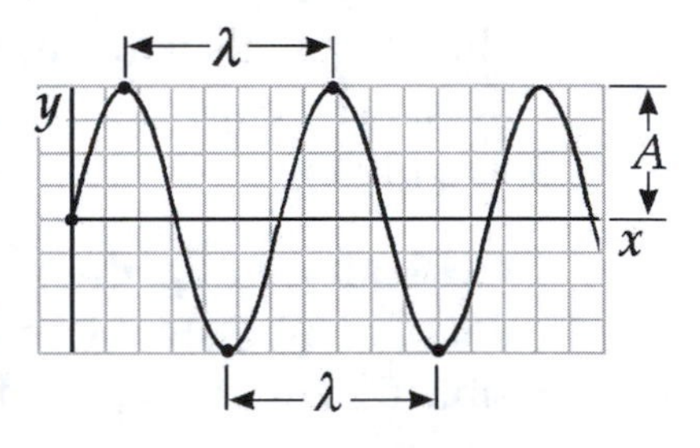

Consider a wave traveling in the x-direction on a very long string. Each particle along the string oscillates in simple harmonic motion along the y-direction with a **frequency** equal to the frequency of the source producing the vibration. The maximum distance the string is displaced above or below the equilibrium value is called the **amplitude**, A, of the wave. The distance between two successive points along the string that are the same distance from equilibrium and moving in the same direction is called the **wavelength**, λ. The number of crests or peaks of the wave passing a fixed point each second is the wave frequency (or harmonic frequency). The wave will advance along the string a distance of one wavelength in a time interval equal to one **period** of vibration, T; the time required for any point in the medium to complete one full cycle of vibration.

13.9 The Speed of Waves on Strings

For linear waves, the **speed of mechanical waves** depends only on the physical properties of the medium through which the disturbance travels. In the case of **waves on a string**, the velocity depends on the tension in the string and its mass per unit length (linear mass density).

13.10 Interference of Waves

If two or more waves are moving through a medium, the **resultant wave function** is the algebraic sum of the wave functions of the individual waves. Two traveling waves can pass through each other without being destroyed or altered.

13.11 Reflection of Waves

Whenever a traveling wave reaches a boundary, **reflection** occurs; part or all of the wave is reflected back into the original medium. If the wave is traveling along a string and is reflected from a "fixed" end, the reflected pulse is inverted. By contrast, a pulse is reflected without inversion at the "free" end of a string.

EQUATIONS AND CONCEPTS

Hooke's law gives the force exerted by a spring that is stretched or compressed beyond the equilibrium position. The force constant, k, is always positive and has a value which corresponds to the relative stiffness of the spring. *The* negative sign in Equation 13.1 indicates *that the force exerted by the string on the mass is always directed opposite the displacement toward the equilibrium position* (the vertical dashed line at $x = 0$).

$$F_s = -kx \qquad (13.1)$$

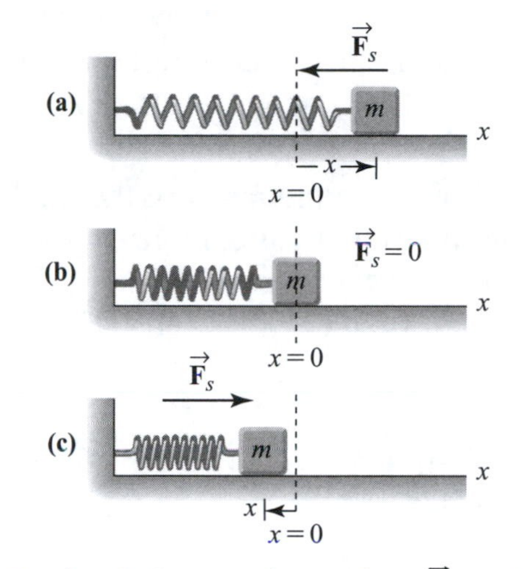

In simple harmonic motion, $\vec{\mathbf{F}} = -k\vec{\mathbf{x}}$.

Acceleration of a mass in simple harmonic motion has a magnitude proportional to the displacement and oppositely directed. *Note that the acceleration of a harmonic oscillator does not remain constant during the motion.* The acceleration equals zero when the oscillating mass is at the equilibrium position ($x = 0$) and has its maximum magnitude when $x = \pm A$ (the amplitude).

$$a = -\left(\frac{k}{m}\right)x \qquad (13.2)$$

Acceleration as a function of position

Elastic potential energy is stored in a spring as a result of work done by an external force to stretch or compress the spring.

$$PE_s \equiv \tfrac{1}{2}kx^2 \qquad (13.3)$$

Conservation of mechanical energy in a spring-mass system (including kinetic energy, gravitational potential energy, and elastic potential energy) holds when only conservative forces are acting on the system.

$$(KE + PE_g + PE_s)_i = (KE + PE_g + PE_s)_f \qquad (13.4)$$

The **speed of an object in simple harmonic motion** is a maximum at $x = 0$; the speed is zero when the mass is at the points of maximum displacement ($x = \pm A$). *This expression can be found from the principle of conservation of mechanical energy.*

$$v = \pm\sqrt{\frac{k}{m}(A^2 - x^2)} \quad (13.6)$$

Speed as a function of position

The **period of a mass-spring system** depends on the mass (an inertial quantity) and the spring constant (an elastic quantity). The period of any object in simple harmonic motion is the time required to complete one full cycle of its motion.

$$T = 2\pi\sqrt{\frac{m}{k}} \quad (13.8)$$

The **harmonic frequency** (f) is the number of cycles or oscillations completed per unit time and is the reciprocal of the period. The units of frequency are hertz (Hz).

$$f = \frac{1}{T} \quad (13.9)$$

$$f = \frac{1}{2\pi}\sqrt{\frac{k}{m}} \quad (13.10)$$

Angular frequency (ω) is measured in radians per second.

$$\omega = 2\pi f = \sqrt{\frac{k}{m}} \quad (13.11)$$

The **initial conditions** illustrated in the figure are required for the particular form of Equations 13.14a, 13.14b, and 13.14c as stated below. Note that the mass is at rest ($v = 0$) and located at $x = A$ (the maximum displacement) when $t = 0$.

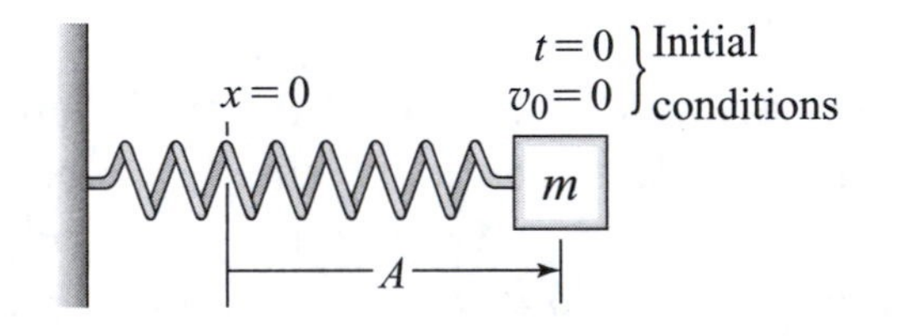

Position as a function of time

$$x = A\cos(2\pi ft) \quad (13.14a)$$

Velocity as a function of time

$$v = -A\omega\sin(2\pi ft) \quad (13.14b)$$

Acceleration as a function of time

$$a = -A\omega^2\cos(2\pi ft) \quad (13.14c)$$

The **period of a simple pendulum** depends only on its length and the local value of the acceleration due to gravity. To a good approximation, the period is independent of the amplitude (θ) within the range of small angular displacements.

$$T = 2\pi\sqrt{\frac{L}{g}} \quad (13.15)$$

The **wave speed**, v, is the rate at which a disturbance or pulse moves along the direction of travel of the wave. *A traveling wave advances a distance equal to one wavelength in a time interval equal to one period.*

$$v = f\lambda \tag{13.17}$$

The **wave speed in a stretched string** depends on the tension in the string and its linear density (mass per unit length). *For any mechanical wave, the speed depends only on the properties of the medium through which the wave travels.*

$$v = \sqrt{\frac{F}{\mu}} \tag{13.18}$$

F = Tension in the string

REVIEW CHECKLIST

- Describe the general characteristics of a system in simple harmonic motion; define amplitude, period, frequency, and displacement.
- Define the following terms relating to wave motion: frequency, wavelength, velocity, and amplitude; express a given harmonic wave function in several alternative forms involving different combinations of the wave parameters: wavelength, period, phase velocity, angular frequency, and harmonic frequency.
- Given a specific wave function for a harmonic wave, obtain values for the characteristic wave parameters: A, ω, and f.
- Make calculations which involve the relationships between wave speed and the inertial and elastic characteristics of a string through which the disturbance is propagating.
- Define and describe the following wave-associated phenomena: superposition, phase, interference, and reflection.

SOLUTIONS TO SELECTED END-OF-CHAPTER PROBLEMS

1. A block of mass $m = 0.60$ kg attached to a spring with force constant 130 N/m is free to move on a frictionless, horizontal surface as in Figure P13.1. The block is released from rest after the spring is stretched a distance $A = 0.13$ m. At that instant, find (a) the force on the block and (b) its acceleration.

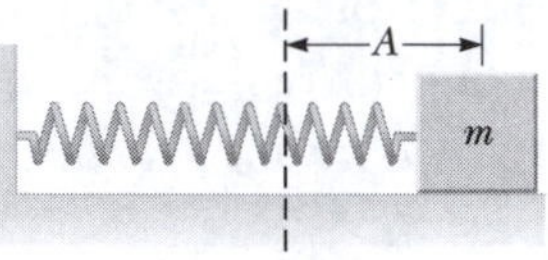

Figure P13.1

Solution

(a) Assuming that the spring obeys Hooke's law, the force it exerts on the block when the spring is stretched 0.13 m beyond its equilibrium length is

$$F_s = -kx = -(130 \text{ N/m})(+0.13 \text{ m}) = -17 \text{ N}$$

or

$F_s = 17$ N directed to the left (i.e., back toward the equilibrium position). ◊

(b) The acceleration given the block by the spring force at the instant the block is released from rest will be

$$a = \frac{F_s}{m} = \frac{-17 \text{ N}}{0.60 \text{ kg}} = -28 \text{ m/s}^2$$

or

$a = 28 \text{ m/s}^2$ directed to the left (again, toward the equilibrium position.) ◊

---

7. A spring 1.50 m long with force constant 475 N/m is hung from the ceiling of an elevator, and a block of mass 10.0 kg is attached to the bottom of the spring. (a) By how much is the spring stretched when the block is slowly lowered to its equilibrium point? (b) If the elevator subsequently accelerates upward at 2.00 m/s^2, what is the position of the block, taking the equilibrium position found in part (a) as $y = 0$ and upwards as the positive y-direction. (c) If the elevator cable snaps during the acceleration, describe the subsequent motion of the block relative to the freely falling elevator. What is the amplitude of its motion?

Solution

(a) We assume the spring force obeys Hooke's law, $F_s = -kx$, where x is the displacement from the position of the lower end of the unstretched spring. The negative sign indicates this force is always directed opposite to the displacement x. When the object comes to equilibrium with $x = -x_0$ (upward taken as positive), Newton's second law gives $\Sigma F_y = +kx_0 - mg = 0$. Thus,

$$x_0 = \frac{mg}{k} = \frac{(10.0 \text{ kg})(9.80 \text{ m/s}^2)}{475 \text{ N/m}} = 0.206 \text{ m}$$

or, with the elevator not accelerating, the equilibrium position is located 0.206 m below the unstretched position of the lower end of the spring. ◊

(b) When the elevator and its contents accelerate upward with $a_y = +2.00\ \text{m/s}^2$, the total displacement of the end of the spring is $x = -x_0 + y$, where $y = 0$ at the equilibrium position found in part (a) above. Then, Newton's second law gives $\Sigma F_y = +kx_0 - ky - mg = ma_y$, or since $kx_0 - mg = 0$ (see above), $-ky = ma_y$ and

$$y = -\frac{ma_y}{k} = -\frac{(10.0\ \text{kg})(+2.00\ \text{m/s}^2)}{475\ \text{N/m}} = -4.21\times10^{-2}\ \text{m} = -4.21\ \text{cm}$$

Thus, the block is located 4.21 cm below the original equilibrium position. ◊

(c) If the cable snaps and the elevator goes into free-fall ($a_y = -g$), our Newton's second law equation becomes $\Sigma F_y = +kx_0 - ky - \cancel{mg} = \cancel{m(-g)}$, which gives $y = +x_0$. Thus, the new "equilibrium position" is the unstretched location of the lower end of the spring. When the cable snapped, the block was located $0.206\ \text{m} + 4.21\ \text{cm} = 0.248\ \text{m}$ below this new "equilibrium" position. While the elevator is in free-fall, the block will oscillate about the unstretched position of the lower end of the spring with amplitude $A = 0.248\ \text{m}$. ◊

---

13. A 10.0-g bullet is fired into, and embeds itself in, a 2.00-kg block attached to a spring with a force constant of 19.6 N/m and whose mass is negligible. How far is the spring compressed if the bullet has a speed of 300 m/s just before it strikes the block and the block slides on a frictionless surface? [*Note:* You must use conservation of momentum in this problem because of the inelastic collision between the bullet and block.]

Solution

The impact between the bullet and the block is an inelastic collision, and it is very difficult to account for the various forms of energy present during the impact. This means that it is not practical to apply conservation of energy to any time interval that includes the moment of impact. We avoid this difficulty by using conservation of momentum from the instant just before the bullet strikes the block to the instant *immediately after impact* (i.e., before the block has had time to move and begin to compress the spring). If v_0 is the speed of the bullet before impact and v_i is the speed of the block (with embedded bullet) immediately after impact, this gives

$$(m_{\text{block}} + m_{\text{bullet}})v_i = m_{\text{bullet}}v_0$$

or

$$v_i = \frac{m_{\text{bullet}}v_0}{m_{\text{block}} + m_{\text{bullet}}} = \frac{(10.0\times10^{-3}\ \text{kg})(300\ \text{m/s})}{2.00\ \text{kg} + 10.0\times10^{-3}\ \text{kg}} = 1.49\ \text{m/s}$$

Now, we apply conservation of energy from immediately after impact until the block comes to rest, obtaining (with $y_f = y_i$ since the surface is horizontal)

$$KE_f + PE_{g,f} + PE_{s,f} = KE_i + PE_{g,i} + PE_{s,i}$$

or

$$0 + \cancel{(m_{\text{block}} + m_{\text{bullet}})gy_f} + \frac{1}{2}kx_f^2 = \frac{1}{2}(m_{\text{block}} + m_{\text{bullet}})v_i^2 + \cancel{(m_{\text{block}} + m_{\text{bullet}})gy_i} + 0$$

and giving the final distance the spring is compressed as

$$x_f = v_i\sqrt{\frac{m_{\text{block}} + m_{\text{bullet}}}{k}} = (1.49 \text{ m/s})\sqrt{\frac{2.01 \text{ kg}}{19.6 \text{ N/m}}} = 0.477 \text{ m} \quad \lozenge$$

---

21. A horizontal spring attached to a wall has a force constant of $k = 850$ N/m. A block of mass $m = 1.00$ kg is attached to the spring and rests on a frictionless, horizontal surface as in Figure P13.21. (a) The block is pulled to a position $x_i = 6.00$ cm from equilibrium and released. Find the potential energy stored in the spring when the block is 6.00 cm from equilibrium. (b) Find the speed of the block as it passes through the equilibrium position. (c) What is the speed of the block when it is at a position $x_i/2 = 3.00$ cm?

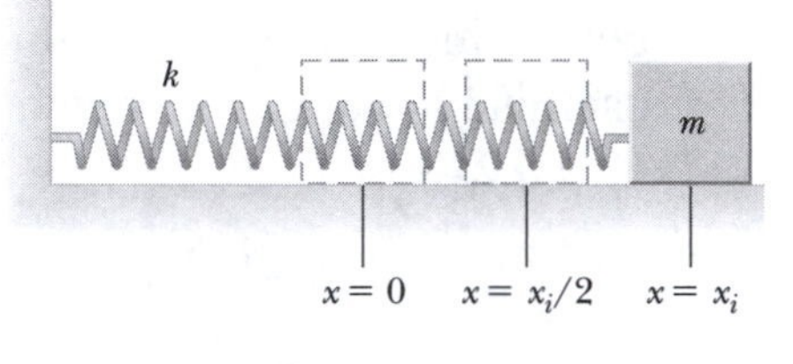

Figure P13.21

Solution

(a) When a spring with force constant k is stretched or compressed by distance x from its normal, relaxed length, the elastic potential energy stored in that spring is given by $PE_s = \frac{1}{2}kx^2$. Thus, the stored energy in the spring when the block is released from rest at $x_i = 6.00$ cm is

$$PE_s = \frac{1}{2}kx_i^2 = \frac{1}{2}(850 \text{ N/m})(6.00 \times 10^{-2} \text{ m})^2 = 1.53 \text{ J} \quad \lozenge$$

(b) As the block moves horizontally, the only force doing work on it is the spring force, a conservative force. Hence, we apply conservation of energy to the block-spring system from the instant the block is released from rest at $x_i = 6.00$ cm until it reaches the equilibrium position $(x_f = 0)$. Since the surface is horizontal, $y_f = y_i$, and $(KE + PE_g + PE_s)_f = (KE + PE_g + PE_s)_i$ becomes $\frac{1}{2}mv_f^2 + \cancel{mgy_f} + 0 = 0 + \cancel{mgy_i} + \frac{1}{2}kx_i^2$, yielding

$$v_f = \sqrt{\frac{kx_i^2}{m}} = \sqrt{\frac{(850 \text{ N/m})(6.00 \times 10^{-2} \text{ m})^2}{1.00 \text{ kg}}} = 1.75 \text{ m/s} \quad \lozenge$$

(c) Again, we apply conservation of energy, this time with $x_f = 3.00$ cm, to obtain $\frac{1}{2}mv_f^2 + \cancel{mgy_f} + \frac{1}{2}kx_f^2 = 0 + \cancel{mgy_i} + \frac{1}{2}kx_i^2$ which gives

$$v_f = \sqrt{\frac{k(x_i^2 - x_f^2)}{m}} = \sqrt{\frac{(850 \text{ N/m})\left[(6.00 \times 10^{-2})^2 - (3.00 \times 10^{-2})^2\right]\text{m}^2}{1.00 \text{ kg}}} = 1.51 \text{ m/s} \quad \lozenge$$

---

27. A cart of mass 250 g is placed on a frictionless horizontal air track. A spring having a spring constant of 9.5 N/m is attached between the cart and the left end of the track. If the cart is displaced 4.5 cm from its equilibrium position, find (a) the period at which it oscillates, (b) its maximum speed, and (c) its speed when it is located 2.0 cm from its equilibrium position.

Solution

(a) The period of an oscillating spring-object system is given by

$$T = \frac{2\pi}{\omega} = \frac{2\pi}{\sqrt{k/m}} = 2\pi\sqrt{\frac{m}{k}}$$

Thus, the period of this spring-cart system will be

$$T = 2\pi\sqrt{\frac{0.250\text{ kg}}{9.5\text{ N/m}}} = 1.0\text{ s}$$ ◊

(b) Since the cart is released from rest 4.5 cm from the equilibrium position, it will undergo simple harmonic motion with amplitude $A = 4.5$ cm about the equilibrium position. At the turning points ($x = \pm A$), all of the energy is in the form of elastic potential energy, while at the equilibrium position ($x = 0$), all of the energy is in the form of kinetic energy. Hence, $E = \frac{1}{2}mv_{\text{max}}^2 = \frac{1}{2}kA^2$, giving

$$v_{\text{max}} = A\sqrt{\frac{k}{m}} = \left(4.5\times10^{-2}\right)\sqrt{\frac{9.5\text{ N/m}}{0.250\text{ kg}}} = 0.28\text{ m/s}$$ ◊

(c) When the cart has a displacement of $x = 2.0$ cm from the equilibrium position, conservation of energy, ignoring the constant gravitation potential energy of the cart on this level track, gives $E = KE + PE_s = 0 + PE_{s,\text{turning point}}$, or

$$\frac{1}{2}mv^2 + \frac{1}{2}kx^2 = 0 + \frac{1}{2}kA^2$$

and

$$v = \sqrt{\left(\frac{k}{m}\right)\left(A^2 - x^2\right)} = \sqrt{\left(\frac{9.5\text{ N/m}}{0.250\text{ kg}}\right)\left[(0.045\text{ m})^2 - (0.020\text{ m})^2\right]} = 0.25\text{ m/s}$$ ◊

31. A 2.00-kg object on a frictionless horizontal track is attached to the end of a horizontal spring whose force constant is 5.00 N/m. The object is displaced 3.00 m to the right from its equilibrium position and then released, initiating simple harmonic motion. (a) What is the force (magnitude and direction) acting on the object 3.50 s after it is released? (b) How many times does the object oscillate in 3.50 s?

Solution

An object attached to a spring on a frictionless horizontal track moves under the influence of a Hooke's law force, $F_s = -kx$. Thus, it will execute simple harmonic motion with an angular frequency

$$\omega = \sqrt{k/m} = \sqrt{(5.00\ \text{N/m})/(2.00\ \text{kg})} = 1.58\ \text{rad/s}$$

Taking to the right as the positive direction, the object starts from rest at $t = 0$ with a displacement of $x = +3.00$ m from the equilibrium position. Therefore, the displacement is a cosine function of time, $x = A\cos(\omega t)$, with $A = 3.00$ m.

(a) At $t = 3.50$ s, $\omega t = (1.58\ \text{rad/s})(3.50\ \text{s}) = 5.53$ rad. Thus, we must set our calculator to operate in the radians mode, not degrees, and find the displacement at this time to be

$$x = A\cos(\omega t) = (3.00\ \text{m})\cos(5.53\ \text{rad}) = +2.19\ \text{m}$$

The force exerted on the object by the spring at this time will be

$$F_s = -kx = -(5.00\ \text{N/m})(+2.19\ \text{m}) = -11.0\ \text{N}$$

or $F_s = 11.0$ N directed toward the left. ◊

(b) The time required for one complete oscillation (or the period) of the simple harmonic motion this object carries out is $T = 2\pi/\omega = 2\pi\sqrt{m/k}$. Hence, the number of oscillations the object has made in the first 3.50 s of motion is given by

$$n = \frac{\Delta t}{T} = \frac{\Delta t}{2\pi\sqrt{m/k}} = \frac{\Delta t}{2\pi}\sqrt{\frac{k}{m}} = \frac{3.50\ \text{s}}{2\pi}\sqrt{\frac{5.00\ \text{N/m}}{2.00\ \text{kg}}} = 0.881$$ ◊

37. A pendulum clock that works perfectly on Earth is taken to the Moon. (a) Does it run fast or slow there? (b) If the clock is started at 12:00 midnight, what will it read after one Earth day (24.0 h)? Assume that the free-fall acceleration on the Moon is $1.63\ \text{m/s}^2$.

Solution

(a) The period of a simple pendulum is $T = 2\pi\sqrt{\ell/g}$, where ℓ is the length of the pendulum and g is the local free-fall acceleration. Thus, if the pendulum clock is taken to the Moon, where the free-fall acceleration is less than on Earth, the pendulum will have a longer period, causing the clock to run slow. ◊

(b) The ratio of the pendulum's period on the Moon to the period it had on Earth is

$$\frac{T_{\text{Moon}}}{T_{\text{Earth}}} = \frac{2\pi\sqrt{\ell/g_{\text{Moon}}}}{2\pi\sqrt{\ell/g_{\text{Earth}}}} = \sqrt{\frac{g_{\text{Earth}}}{g_{\text{Moon}}}}$$

or

$$T_{\text{Moon}} = T_{\text{Earth}}\sqrt{\frac{g_{\text{Earth}}}{g_{\text{Moon}}}}$$

Since the frequency of the pendulum is the reciprocal of its period, the frequency of the clock's "ticks" on the Moon is

$$f_{\text{Moon}} = \frac{1}{T_{\text{Moon}}} = \frac{1}{T_{\text{Earth}}\sqrt{g_{\text{Earth}}/g_{\text{Moon}}}} = f_{\text{Earth}}\sqrt{\frac{g_{\text{Moon}}}{g_{\text{Earth}}}}$$

This means that when a time $(\Delta t)_{\text{Earth}}$ has elapsed on Earth clocks, the time that will appear to have passed according to the pendulum clock on the Moon is $(\Delta t)_{\text{Moon}} = (\Delta t)_{\text{Earth}}\sqrt{g_{\text{Moon}}/g_{\text{Earth}}}$. If the elapsed time on Earth is $(\Delta t)_{\text{Earth}} = 24.0$ h, the time that has passed according to the moon-based clock is

$$(\Delta t)_{\text{Moon}} = 24.0\ \text{h}\sqrt{\frac{1.63\ \text{m/s}^2}{9.80\ \text{m/s}^2}} = 9.79\ \text{h} = 9\ \text{h} + (0.79\ \text{h})\left(\frac{60\ \text{min}}{1\ \text{h}}\right) = 9\ \text{h}\ + 47\ \text{min}$$

Since the pendulum clock on the Moon started with a reading of 12:00 midnight, its current reading will be 9:47 AM. ◊

42. An object attached to a spring vibrates with simple harmonic motion as described by Figure P13.42. For this motion, find (a) the amplitude, (b) the period, (c) the angular frequency, (d) the maximum speed, (e) the maximum acceleration, and (f) an equation for its position x in terms of a sine function.

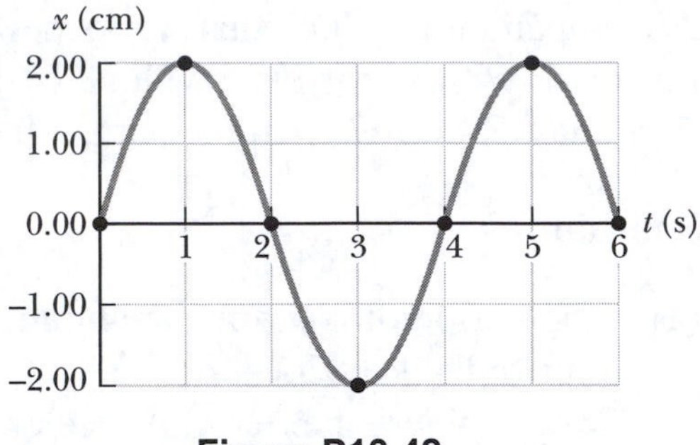

Figure P13.42

Solution

(a) The amplitude of a simple harmonic motion is the magnitude of the maximum displacement from the equilibrium position ($x = 0$). From the graph of Figure P13.42, we see that the amplitude of this motion is $A = 2.00$ cm. ◊

(b) The period of a simple harmonic motion is the time required to complete one full cycle of the oscillation. For the motion described by Figure P13.42, it is seen that the period is $T = 4.00$ s. ◊

(c) The period and the angular frequency of a simple harmonic motion are related by the expression $T = 2\pi/\omega$. Thus, the angular frequency of the motion shown in Figure P13.42 is $\omega = 2\pi/T = 2\pi/(4.00\text{ s}) = (\pi/2)\text{ rad/s}$. ◊

(d) The constant total energy of an undamped harmonic oscillator is equal to the elastic potential energy, $PE_s = \frac{1}{2}kx^2$, at the turning points ($x = \pm A$), which is the same as the kinetic energy at the equilibrium position, $KE_{\text{max}} = \frac{1}{2}mv_{\text{max}}^2$. That is, $E = \frac{1}{2}mv_{\text{max}}^2 = \frac{1}{2}kA^2$, so the maximum speed is $v_{\text{max}} = A\sqrt{k/m} = A\omega$, where $\omega = \sqrt{k/m}$ is the angular frequency. For the motion described in Figure P13.42,

$$v_{\text{max}} = A\omega = (2.00\text{ cm})\left(\frac{\pi}{2}\text{ rad/s}\right) = \pi\text{ cm/s}$$ ◊

(e) Simple harmonic motion is the result of a Hooke's law restoring force, $F_s = -kx$. Thus, the magnitude of the maximum force exerted on the object is given by $|F_s|_{\text{max}} = k|x_{\text{max}}| = kA$, and the maximum acceleration produced by this force has magnitude $a_{\text{max}} = |F_s|_{\text{max}}/m = kA/m = A(k/m) = A\omega^2$. In this case, we have $a_{\text{max}} = (2.00\text{ cm})[(\pi/2)\text{ rad/s}]^2 = 4.93\text{ cm/s}^2$. ◊

(f) Simple harmonic motion is a sinusoidal function of time. Since, in this case, the displacement is $x = 0$ at $t = 0$, we must use the sine function giving $x = A\sin(\omega t)$, or $x = (2.00\text{ cm})\sin(\pi t/2)$. ◊

53. Transverse waves with a speed of $50.0\ \text{m/s}$ are to be produced on a stretched string. A 5.00-m length of string with a total mass of 0.060 0 kg is used. (a) What is the required tension in the string? (b) Calculate the wave speed in the string if the tension is 8.00 N.

Solution

The mass per unit length of this string is

$$\mu = \frac{m}{\ell} = \frac{0.060\,0\ \text{kg}}{5.00\ \text{m}} = 1.20\times10^{-2}\ \text{kg/m}$$

(a) The speed of transverse waves in a stretched string is given by (see Equation 13.18 in the textbook)

$$v = \sqrt{\frac{F}{\mu}}$$

where F is the tension in the string and μ is the mass per unit length of the string. Therefore, if the wave speed is to be $50.0\ \text{m/s}$ in the given string, the required tension in the string is

$$F = \mu v^2 = \left(1.20\times10^{-2}\ \text{kg/m}\right)\left(50.0\ \text{m/s}\right)^2 = 30.0\ \text{N}$$ ◊

(b) If the tension in the string is now $F = 8.00$ N, the current speed of transverse waves in the string is

$$v = \sqrt{\frac{F}{\mu}} = \sqrt{\frac{8.00\ \text{N}}{1.20\times10^{-2}\ \text{kg/m}}} = 25.8\ \text{m/s}$$ ◊

58. The elastic limit of a piece of steel wire is 2.70×10^9 Pa. What is the maximum speed at which transverse wave pulses can propagate along the wire without exceeding its elastic limit? (The density of steel is 7.86×10^3 kg/m^3.)

Solution

When a wire has tension F in it, the tensile stress is $Stress = F/A$, where A is the cross-sectional area of the wire. If the mass per unit length of the wire is $\mu = m/L$, the speed of transverse waves in the wire may be written as

$$v = \sqrt{\frac{F}{\mu}} = \sqrt{\frac{A \cdot Stress}{m/L}} = \sqrt{\frac{(A \cdot L)Stress}{m}}$$

The product $A \cdot L$, where A is the wire's cross-sectional area and L is its length, is simply the volume V of material within the wire. Also, the mass of the wire is given by $m = \rho V$, where ρ is the density of the material making up the wire. The speed of transverse waves in the wire then becomes

$$v = \sqrt{\frac{(V)Stress}{\rho V}} = \sqrt{\frac{Stress}{\rho}}$$

If the maximum stress that can exist in a wire of density $\rho = 7.86 \times 10^3$ kg/m^3 is $(Stress)_{max} = 2.70 \times 10^9$ Pa, the maximum speed of transverse waves in this wire is

$$v_{max} = \sqrt{\frac{(Stress)_{max}}{\rho}} = \sqrt{\frac{2.70 \times 10^9 \text{ Pa}}{7.86 \times 10^3 \text{ kg/m}^3}} = 586 \text{ m/s}$$ ◊

62. The position of a 0.30-kg object attached to a spring is described by

$$x = (0.25 \text{ cm})\cos(0.4\pi t)$$

Find (a) the amplitude of the motion, (b) the spring constant, (c) the position of the object at $t = 0.30$ s, and (d) the object's speed at $t = 0.30$ s.

Solution

(a) The displacement of the given object from the equilibrium position varies sinusoidally in time as $x = (0.25 \text{ cm})\cos(0.4\pi t)$. Compare this function to the general equation $x = A\cos(\omega t)$ for the displacement of an object that is executing simple harmonic motion and has maximum displacement at time $t = 0$. We conclude that the given object undergoes simple harmonic motion with an amplitude (maximum displacement) of $A = 0.25$ m and angular frequency $\omega = 0.4\pi$ rad/s. ◊

(b) The angular frequency of an object executing simple harmonic motion is determined by the spring constant k and the mass m according to the relation $\omega = \sqrt{k/m}$. Thus, the spring constant is $k = m\omega^2$, and in this case its value is

$$k = m\omega^2 = (0.30 \text{ kg})(0.4\pi \text{ rad/s})^2 = 0.47 \text{ kg/s}^2 = 0.47 \text{ N/m}$$ ◊

(c) At $t = 0.30$ s, the argument of the cosine function is

$$\omega t = (0.4\pi \text{ rad/s})(0.30 \text{ s}) = 0.12\pi \text{ rad}$$

Thus, to compute the position of the object at this time, we must place our calculator in radians mode (not degrees) and evaluate

$$x = (0.25 \text{ m})\cos(0.12\pi \text{ rad}) = 0.23 \text{ m}$$

(d) When an object attached to a spring is at one of its turning points ($x = \pm A$), all of the energy of the system is stored as elastic potential energy $PE_s = \frac{1}{2}kx^2$, and the total energy is given by $E = \frac{1}{2}kA^2$. If the object is not at one of the turning points, the constant total energy of the system is divided between potential energy and kinetic energy. That is, $\frac{1}{2}mv^2 + \frac{1}{2}kx^2 = \frac{1}{2}kA^2$, and the speed of the object is $v = \sqrt{k\left(A^2 - x^2\right)/m} = \sqrt{\omega^2\left(A^2 - x^2\right)} = \omega\sqrt{A^2 - x^2}$. Thus at $t = 0.30$ s,

$$v = (0.4\pi \text{ rad/s})\sqrt{(0.25 \text{ m})^2 - (0.23 \text{ m})^2} = 0.12 \text{ m/s}$$ ◊

65. A simple pendulum has mass 1.20 kg and length 0.700 m. (a) What is the period of the pendulum near the surface of Earth? (b) If the same mass is attached to a spring, what spring constant would result in the period of motion found in part (a)?

Solution

(a) The motion of a simple pendulum, swinging with small amplitude oscillations, very closely approximates simple harmonic motion. The period of this motion is given by $T = 2\pi\sqrt{\ell/g}$, where ℓ is the length of the pendulum and g is the local free-fall acceleration. If the pendulum is located on the surface of Earth (assumed near sea level), the free-fall acceleration is $g = 9.80\ \text{m/s}^2$, and the period of a simple pendulum having a length of $\ell = 0.700$ m is

$$T = 2\pi\sqrt{\frac{\ell}{g}} = 2\pi\sqrt{\frac{0.700\ \text{m}}{9.80\ \text{m/s}^2}} = 1.68\ \text{s} \qquad \Diamond$$

(b) A mass m attached to the end of a spring, having spring constant k, executes simple harmonic motion with an angular frequency of $\omega = \sqrt{k/m}$ and period $T = 2\pi/\omega = 2\pi\sqrt{m/k}$. Thus, if $m = 1.20$ kg and the period is to be the same as that found in part (a), the required spring constant is

$$k = \frac{4\pi^2 m}{T^2} = \frac{4\pi^2(1.20\ \text{kg})}{(1.68\ \text{s})^2} = 16.8\ \ \text{kg/s}^2 = 16.8\ \text{N/m} \qquad \Diamond$$

14

Sound

NOTES FROM SELECTED CHAPTER SECTIONS

14.1 Producing a Sound Wave
14.2 Characteristics of Sound Waves

Sound waves, which have as their source vibrating objects, are **longitudinal waves** traveling through a medium such as air. The particles of the medium oscillate back and forth along the direction in which the wave travels. This is in contrast to a transverse wave, in which the vibrations of the medium are at right angles to the direction of travel of the wave.

A sound wave traveling through air creates alternating regions of high and low molecular density and air pressure. A region of high density and air pressure is called a **compression** or **condensation**; a region of lower-than-normal density is referred to as a **rarefaction**. A sinusoidal curve can be used to represent a sound wave. *There are crests in the sinusoidal wave at the points where the sound wave has condensations, and troughs where the sound wave has rarefactions.*

14.4 Energy and Intensity of Sound Waves

The **intensity** of a wave is the rate at which sound energy flows through a unit area perpendicular to the direction of travel of the wave. The faintest sounds the human ear can detect have an intensity of about 1×10^{-12} W/m^2. This intensity is called the **threshold of hearing**. The loudest sounds the ear can tolerate, at the **threshold of pain**, have an intensity of about 1 W/m^2. The sensation of loudness is approximately logarithmic in the human ear, and the relative intensity of a sound is called the **intensity level** or **decibel level**.

14.5 Spherical and Plane Waves

The **intensity** of a spherical wave produced by a point source is proportional to the average power emitted and inversely proportional to the square of the distance from the source.

14.6 The Doppler Effect

In general, the Doppler effect is experienced whenever there is relative motion between source and observer. When the source and observer are moving toward each other, the frequency heard by the observer is higher than the frequency of the source. When the source and observer are moving away from each other, the observer hears a frequency lower than the source frequency.

14.8 Standing Waves

Standing waves can be set up in a string by the superposition of two wave trains traveling in opposite directions along the string. This can occur when waves reflected from one end of the string interfere with the incident wave train, under the condition that the wavelengths (and therefore frequencies) are properly matched to the length of the string. The string has a number of natural patterns of vibration, called **normal modes**. Each normal mode has a **characteristic frequency**. The lowest of these frequencies is called the **fundamental frequency**, which together with the higher frequencies form a **harmonic series**.

The figure on the right is a schematic representation of standing waves in a string of length L. In each case the envelope represents successive positions of the string during one complete cycle. Imagine that you observe the string vibrate while it is illuminated with a strobe light. The first three normal modes are shown. The points of zero displacement are called **nodes** (identified by N); the points of maximum displacement are called **antinodes** (identified by A).

14.9 Forced Vibrations and Resonance

Consider a system (for example a mass-spring system) that has a natural frequency of vibration, f_o, and is driven or pushed back and forth with a periodic force whose frequency is f. This type of motion is referred to as a **forced vibration**. Its amplitude reaches a maximum when the frequency of the driving force equals the natural frequency of the system, f_o, called the **resonant frequency** of the system. Under this condition, the system is said to be in **resonance**.

14.10 Standing Waves in Air Columns

Sound sources can be used to produce **longitudinal standing waves** in air columns. The phase relationship between incident and reflected waves depends on whether or not the reflecting end of the air column is open or closed. This gives rise to two sets of possible standing wave conditions:

> In a **pipe open at both ends, all harmonics are present**; the natural frequencies of vibration form a series with frequencies equal to integral multiples of the fundamental.
>
> In a **pipe closed at one end and open at the other, only odd harmonics are present**.

14.11 Beats

Consider a type of interference effect that results from the superposition of two waves with slightly different frequencies. In this situation, at some fixed point the waves are periodically in and out of phase, resulting in alternating periods of constructive and destructive interference. A listener hears an alternation in loudness, known as beats. The number of beats per second, or beat frequency, equals the difference in frequency between the two sources.

EQUATIONS AND CONCEPTS

The **speed of sound in a fluid** depends on the value of the bulk modulus, B (an elastic property), and the equilibrium density, ρ (an inertial property), of the material through which it is traveling.

$$v = \sqrt{\frac{B}{\rho}} \quad (14.1)$$

$$\text{where } B \equiv -\frac{\Delta P}{\Delta V/V} \quad (14.2)$$

The **speed of sound in a solid** depends on the value of Young's modulus and the density of the material.

$$v = \sqrt{\frac{Y}{\rho}} \quad (14.3)$$

The **speed of sound in air** depends on the Kelvin temperature as given by Equation 14.4. The speed of sound in air at 0°C (273 K) is 331 m/s.

$$v = (331 \text{ m/s})\sqrt{\frac{T}{273 \text{ K}}} \quad (14.4)$$

The **intensity of a wave** is the rate at which energy flows across a unit area in a plane perpendicular to the direction of travel of the wave. The SI units of intensity, I, are watts per square meter (W/m^2).

$$I \equiv \frac{\text{power}}{\text{area}} = \frac{P}{A} \quad (14.6)$$

The **decibel scale** is a logarithmic intensity scale. On this scale, the unit of sound intensity is the decibel, dB. The constant I_0 is a reference intensity chosen to coincide with the threshold of hearing. *Note: The term "log" indicates common logarithms (base 10) as opposed to "ln," which indicates natural logarithms (base e).* See an example of calculating the decibel level due to two sources sounded simultaneously in Suggestions, Skills, and Strategies.

$$\beta \equiv 10\log\left(\frac{I}{I_0}\right) \quad (14.7)$$

$$I_0 = 1.0 \times 10^{-12} \text{ W/m}^2$$

The **intensity of a spherical wave produced by a point source** is inversely proportional to the square of the distance from the source. The **ratio of intensities** at spherical surfaces of radii r_1 and r_2 is shown in Equation 14.9.

$$I = \frac{\text{average power}}{\text{area}} = \frac{P_{\text{av}}}{A} = \frac{P_{\text{av}}}{4\pi r^2} \quad (14.8)$$

$$\frac{I_1}{I_2} = \frac{r_2^2}{r_1^2} \quad (14.9)$$

The doppler effect (apparent change in frequency) is observed whenever there is relative motion between the source and the observer. Equation 14.12 is a general Doppler shift expression. **Proper use of signs for v_O and v_S is required depending on the relative motion of source and observer toward or away from each other!** See worked examples in Suggestions, Skills, and Strategies.

$$f_O = f_S\left(\frac{v + v_O}{v - v_S}\right) \qquad (14.12)$$

f_S = frequency of the source

f_O = observed frequency

v = velocity of sound (always positive)

v_O = velocity of observer

v_S = velocity of source

v_O and v_S are each measured relative to the medium in which the sound travels.

Apply the following sign rules when using Equation 14.12.

When the observer moves toward the source, substitute the value of v_O with a positive sign; a negative sign is used when the observer moves away from the source.

When the source moves toward the observer, substitute the value of v_S with a positive sign; a negative sign is used when the source moves away from the observer.

The velocity of sound v is always substituted with a positive sign.

Normal modes of oscillation (a series of natural oscillations or "standing wave patterns") can be excited in a stretched string (fixed at both ends). Each mode corresponds to a characteristic frequency and wavelength.

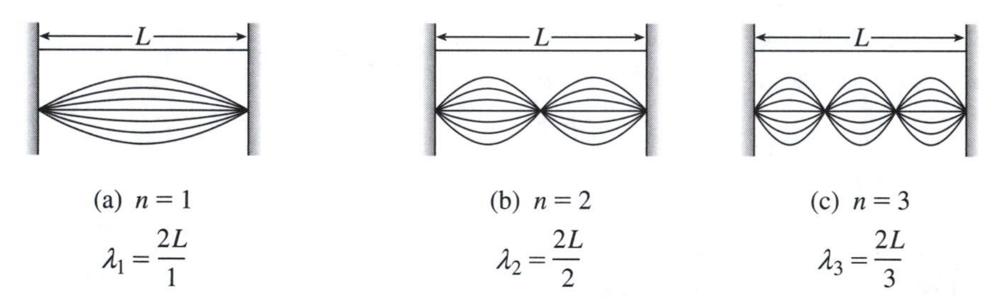

(a) $n = 1$ $\quad \lambda_1 = \frac{2L}{1}$

(b) $n = 2$ $\quad \lambda_2 = \frac{2L}{2}$

(c) $n = 3$ $\quad \lambda_3 = \frac{2L}{3}$

Standing waves in a stretched string of length L fixed at both ends. The normal frequencies of vibration form a harmonic series: (a) the fundamental frequency, or first harmonic, (b) the second harmonic, and (c) the third harmonic.

The **observed frequencies** are integer multiples of the fundamental frequency (or first harmonic) f_1 corresponding to $n = 1$. This is the lowest frequency for which a standing wave is possible. The higher frequencies ($n = 2, 3, 4, \ldots$) form a harmonic series and can be expressed in terms of the wave speed and string length or in terms of the string tension and linear mass-density.

$$f_n = nf_1 = n\frac{v}{2L} \qquad n = 1, 2, 3, \ldots$$

$$f_n = \frac{n}{2L}\sqrt{\frac{F}{\mu}} \qquad n = 1, 2, 3, \ldots \qquad (14.17)$$

F = string tension

μ = linear mass-density

In an **"open" pipe** (open at both ends), the natural frequencies of vibration form a series in which all harmonics (all integer multiples of the fundamental) are present. *Note that this series of frequencies is the same as the series of frequencies produced in a string fixed at both ends.*

$$f_n = n\left(\frac{v}{2L}\right) = nf_1 \quad n = 1, 2, 3, \ldots \quad (14.18)$$

In a **"closed pipe"** (open at one end), only the odd harmonics (odd multiples of the fundamental) are possible.

$$f_n = n\left(\frac{v}{4L}\right) = nf_1 \quad n = 1, 3, 5, \ldots \quad (14.19)$$

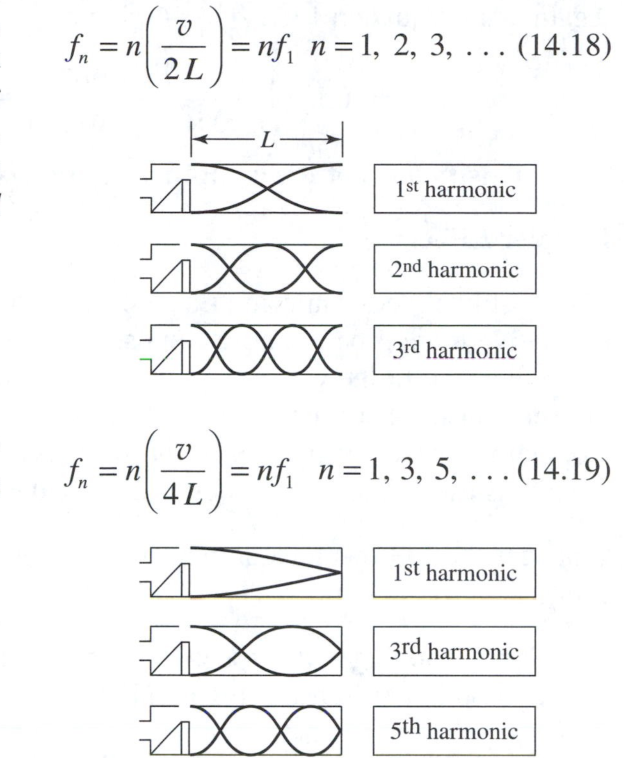

SUGGESTIONS, SKILLS, AND STRATEGIES

The Decibel Scale

When making calculations using Equation 14.7, which defines the intensity level of a sound wave on the decibel scale, the properties of logarithms must be kept clearly in mind.

In order to determine the decibel level corresponding to two sources sounded simultaneously, you must first find the intensity, I, of each source in W/m^2, add these values, and then convert the resulting intensity to the decibel scale.

As an example of this technique, determine the dB level when two sounds with intensities of $\beta_1 = 40.0$ dB and $\beta_2 = 45.0$ dB are sounded together.

First solve Equation 14.7, $\beta = 10 \log (I/I_0)$, to find an expression for the intensity, $I = I_0 10^{\beta/10}$.

Then for $\beta_1 = 40.0$ dB,

$$I_1 = (10^{-12}\ \text{W/m}^2)10^4 = 1.00 \times 10^{-8}\ \text{W/m}^2$$

For $\beta_2 = 45.0$ dB

$$I_2 = (10^{-12}\ \text{W/m}^2)10^{4.5} = 3.16 \times 10^{-8}\ \text{W/m}^2$$

and

$$I_{\text{total}} = I_1 + I_2 = 4.16 \times 10^{-8}\ \text{W/m}^2$$

Again using Equation 14.7, $\beta_{total} = 10 \log (I_{total}/I_0)$. So,

$$\beta_{total} = 10 \log \frac{(4.16 \times 10^{-8}\ \text{W/m}^2)}{(1.00 \times 10^{-12}\ \text{W/m}^2)} = 10 \log (4.16 \times 10^4) = 46.2\ \text{dB}$$

The intensity level of the combined sources is 46.2 dB (not 40 dB + 45 dB = 85 dB).

Doppler Effect

The most likely error in using Equation 14.12 to calculate the Doppler frequency shift due to relative motion between a sound source and an observer is due to using the incorrect algebraic sign for the velocity of either the observer or the source. Remember the following relationship between relative velocity of source and observer and the corresponding Doppler frequency shift: The word **toward** is associated with an **increase** in frequency and the words **away from** are associated with a **decrease** in frequency.

Consider the following examples, given a source frequency of 300 Hz and the speed of sound in air of 343 m/s.

(1) Source moving with a speed of 40 m/s toward a fixed observer. In this case, $v_O = 0$ and $v_S = +40$ m/s. Substituting into Equation 14.12,

$$f_O = \left(\frac{v + v_O}{v - v_S}\right) f_S = \left(\frac{343\ \text{m/s} + 0}{343\ \text{m/s} - 40\ \text{m/s}}\right) (300\ \text{Hz}) = 340\ \text{Hz}$$

(2) Source moving away from a fixed observer with a speed of 40 m/s. In this case, $v_O = 0$ and $v_S = -40$ m/s. Substituting into Equation 14.12,

$$f_O = \left(\frac{v + v_O}{v - v_S}\right) f_S = \left(\frac{343\ \text{m/s} + 0}{343\ \text{m/s} - (-40\ \text{m/s})}\right) (300\ \text{Hz}) = 269\ \text{Hz}$$

(3) Observer moving with speed of 40 m/s toward a fixed source: $v_O = +40$ m/s and $v_S = 0$. Substituting into Equation 14.12,

$$f_O = \left(\frac{v + v_O}{v - v_S}\right) f_S = \left(\frac{343\ \text{m/s} + 40\ \text{m/s}}{343\ \text{m/s} - 0}\right) (300\ \text{Hz}) = 335\ \text{Hz}$$

(4) Observer moving with speed 40 m/s away from a fixed source: $v_O = -40$ m/s and $v_S = 0$. Substituting into Equation 14.12,

$$f_O = \left(\frac{v + v_O}{v - v_S}\right) f_S = \left(\frac{343\ \text{m/s} + (-40\ \text{m/s})}{343\ \text{m/s} - 0}\right) (300\ \text{Hz}) = 265\ \text{Hz}$$

As a check on your calculated value of f_O, remember:

- When the relative motion is source or observer toward the other, $f_O > f_S$.
- When the relative motion is source or observer away from the other, $f_O < f_S$.

REVIEW CHECKLIST

- Describe the harmonic displacement and pressure variation as functions of time and position for a harmonic sound wave.
- Calculate the speed of sound in various media in terms of appropriate elastic properties (these can include bulk modulus, Young's modulus, and the pressure-volume relationships of an ideal gas) and the corresponding inertial properties (usually the mass density).
- Understand the basis of the logarithmic intensity scale (decibel scale) and convert intensity values (given in W/m^2) to loudness levels on the dB scale. Determine the intensity ratio for two sound sources whose decibel levels are known. Calculate the intensity of a point source wave at a given distance from the source. Remember that you must use common logarithms (base 10) in Equation 14.7.
- Describe the various situations under which a Doppler-shifted frequency is produced. Calculate the apparent frequency for a given actual frequency for each of the various possible relative motions between source and observer.
- Describe in both qualitative and quantitative terms the conditions that produce standing waves in a stretched string and in an open or closed air column pipe.

SOLUTIONS TO SELECTED END-OF-CHAPTER PROBLEMS

7. A sound wave propagating in air has a frequency of 4.00 kHz. Calculate the change in wavelength when the wave, initially traveling in a region where $T = 27.0°\text{C}$, enters a region where $T = 10.0°\text{C}$.

Solution

The speed at which sound waves propagate through air depends on the temperature of the air, increasing as the temperature increases. Thus, when the sound crosses the boundary separating the 27°C air from the 10°C air, the speed will decrease. However, the frequency of the sound wave will be unchanged. To understand why this must be true, think of wavefronts coming up to this boundary from the hotter air. The wavefronts do not cease to exist at the boundary, nor do they accumulate at the boundary. Rather, each time a wavefront arrives at the boundary, it crosses into the cooler air and travels, at a reduced speed, away from the boundary. Hence, the frequency at which wavefronts arrive at the boundary (the frequency of the sound in the hotter air) must equal the frequency at which wavefronts leave the boundary (the frequency of the sound in the cooler air). Since the wavefronts travel away from the boundary at a lower speed than they had approaching the boundary, the spacing between successive wavefronts (the wavelength of the sound) will be less in the cooler air than it was in the hotter air.

The speed of sound at 27.0°C is

$$v_{27} = (331\ \text{m/s})\sqrt{1+\frac{T_C}{273}} = (331\ \text{m/s})\sqrt{1+\frac{27.0}{273}} = 347\ \text{m/s}$$

and at 10.0°C this speed is

$$v_{10} = (331\ \text{m/s})\sqrt{1+\frac{T_C}{273}} = (331\ \text{m/s})\sqrt{1+\frac{10.0}{273}} = 337\ \text{m/s}$$

The speed of this sound is also given in terms of the wavelength, λ, and the constant frequency, $f = 4.00$ kHz, by the expression $v = \lambda f$. Therefore, the change in the wavelength as the sound crosses from the hotter air into the cooler air is

$$\Delta\lambda = \lambda_{10} - \lambda_{27} = \frac{v_{10}}{f} - \frac{v_{27}}{f} = \frac{337\ \text{m/s} - 347\ \text{m/s}}{4.00\times10^{3}\ \text{Hz}} = -2.5\times10^{-3}\ \text{m} = -2.5\ \text{mm}$$

That is, the wavelength decreases by 2.5 millimeters as the sound passes into the cooler air. ◊

---

13. A person wears a hearing aid that uniformly increases the intensity level of all audible frequencies of sound by 30.0 dB. The hearing aid picks up sound having a frequency of 250 Hz at an intensity of $3.0\times10^{-11}\ \text{W/m}^2$. What is the intensity delivered to the eardrum?

Solution

The decibel level of sound having intensity I is given by $\beta = 10\log(I/I_0)$, where $I_0 = 1.0\times10^{-12}\ \text{W/m}^2$ is the intensity of sound at the threshold of hearing. If the incident sound has intensity $I_1 = 3.0\times10^{-11}\ \text{W/m}^2$, the sound level reaching the ear drum without the hearing aid would be

$$\beta_1 = 10\log\left(\frac{I_1}{I_0}\right) = 10\log\left(\frac{3.0\times10^{-11}\ \text{W/m}^2}{1.0\times10^{-12}\ \text{W/m}^2}\right) = 14.8\ \text{dB}$$

Since the hearing aid increases the sound level of all audible frequencies by 30.0 dB, the sound level reaching the eardrum with the hearing aid in place is

$$\beta_2 = \beta_1 + 30.0\ \text{dB} = 14.8\ \text{dB} + 30.0\ \text{dB} = 44.8\ \text{dB}$$

and the intensity of this sound is

$$I_2 = I_0\cdot10^{\beta_2/10} = (1.0\times10^{-12}\ \text{W/m}^2)\cdot10^{4.48} = 3.0\times10^{-8}\ \text{W/m}^2$$ ◊

As an alternate approach to this problem, observe using the property of logarithms that $\log A - \log B = \log(A/B)$ allows one to write the difference in the decibel level of two sounds as

$$\beta_2 - \beta_1 = 10\log\left(\frac{I_2}{I_0}\right) - 10\log\left(\frac{I_1}{I_0}\right) = 10\log\left(\frac{I_2}{I_0}\cdot\frac{I_0}{I_1}\right) = 10\log\left(\frac{I_2}{I_1}\right)$$

Since the hearing aid increases the decibel level by 30 dB, we have $\beta_2 - \beta_1 = 30$ dB, and the ratio of the intensities reaching the eardrum with and without the hearing aid in place is

$$\frac{I_2}{I_1} = 10^{(\beta_2 - \beta_1)/10} = 10^{(30\text{ dB})/10} = 10^3$$

giving

$$I_2 = 10^3 \cdot I_1 = 10^3 \left(3.0 \times 10^{-11}\ \text{W/m}^2\right) = 3.0 \times 10^{-8}\ \text{W/m}^2$$ ◊

---

21. Show that the difference in decibel levels β_1 and β_2 of a sound source is related to the ratio of its distances r_1 and r_2 from the receivers by the formula

$$\beta_2 - \beta_1 = 20 \log\left(\frac{r_1}{r_2}\right)$$

Solution

Assuming the sound source radiates equally well in all directions, the wave fronts are spheres centered on the source. The intensity of the sound from such a source varies inversely as the square of the distance from the source, or

$$I = \frac{P_{\text{av}}}{4\pi r^2}$$

where P_{av} is the average power emitted by the source and r is the distance of the observer from the source.

Since the decibel level of a sound of intensity I is defined as $\beta = 10\log\left(I/I_0\right)$, where I_0 is a constant, the difference in the decibel levels at distances r_1 and r_2 from a source is given by

$$\beta_2 - \beta_1 = 10\log\left(\frac{I_2}{I_0}\right) - 10\log\left(\frac{I_1}{I_0}\right) = 10\log\left(\frac{I_2}{I_0} \cdot \frac{I_0}{I_1}\right) = 10\log\left(\frac{I_2}{I_1}\right)$$

where we have used the fact that $\log A - \log B = \log\left(A/B\right)$. In terms of the distances from the source, this becomes

$$\beta_2 - \beta_1 = 10\log\left(\frac{P_{\text{av}}}{4\pi r_2^2} \cdot \frac{4\pi r_1^2}{P_{\text{av}}}\right) = 10\log\left(\frac{r_1^2}{r_2^2}\right) = 10\log\left[\left(\frac{r_1}{r_2}\right)^2\right]$$

Finally, making use of the property of logarithms that $\log A^m = m\log A$, we obtain

$$\beta_2 - \beta_1 = 10\log\left[\left(\frac{r_1}{r_2}\right)^2\right] = 10\left[2\log\left(\frac{r_1}{r_2}\right)\right] \quad \text{or} \quad \beta_2 - \beta_1 = 20\log\left(\frac{r_1}{r_2}\right)$$ ◊

---

26. At rest, a car's horn sounds the note A (440 Hz). The horn is sounded while the car is moving down the street. A bicyclist moving in the same direction with one-third the car's speed hears a frequency of 415 Hz. (a) Is the cyclist ahead of or behind the car? (b) What is the speed of the car?

Solution

(a) In general, when a sound source and the observer are getting closer together (due to the motion of the source and/or the motion of the observer), the observer detects a higher frequency than would be heard if the source and observer were a fixed distance apart. When the source and observer are getting farther apart, the observer hears a lower frequency than would otherwise be detected. In the given case, the observer (bicyclist) detects a lower frequency than that heard by an observer at rest relative to the car. Thus, the distance between the bicyclist and the car is increasing. Since the bicyclist and the car are traveling in the same direction, with the car going faster than the bicyclist, the car would be gaining on the bicyclist with the distance between them decreasing if the bicyclist were in front of the car. Thus, we must conclude that the bicyclist is behind the car, with the car pulling away from him. ◊

(b) When both the source and the observer are in motion relative to the medium through which the sound is propagating, the frequency f_O detected by the observer is $f_O = f_S\left[(v+v_O)/(v-v_S)\right]$, where f_S is the frequency emitted by the source. All velocities are measured relative to the medium through which the sound propagates, with v being the velocity of sound, v_O the velocity of the observer, and v_S is that of the source.

In this case, the bicyclist moves *toward* the source with speed $|v_O| = |v_{\text{car}}|/3$, while the car moves *away* from the observer at speed $|v_S| = |v_{\text{car}}|$. We assume the temperature is 20°C, so the speed of sound in air is $v = 343\ \text{m/s}$. According to the sign convention adopted in the discussion of the General Case in Section 14.6 of the textbook, $v_O = +|v_{\text{car}}|/3$ and $v_S = -|v_{\text{car}}|$. Thus, with $f_S = 440$ Hz and $f_O = 415$ Hz, we have

$$415\ \text{Hz} = (440\ \text{Hz})\left(\frac{343\ \text{m/s} + \left(+|v_{\text{car}}|/3\right)}{343\ \text{m/s} - \left(-|v_{\text{car}}|\right)}\right) = (440\ \text{Hz})\left(\frac{343\ \text{m/s} + |v_{\text{car}}|/3}{343\ \text{m/s} + |v_{\text{car}}|}\right)$$

Solving for the speed of the car, we find $|v_{\text{car}}| = 32.0\ \text{m/s}$. ◊

29. A tuning fork vibrating at 512 Hz falls from rest and accelerates at $9.80\ \text{m/s}^2$. How far below the point of release is the tuning fork when waves of frequency 485 Hz reach the release point?

Solution

With $v_S = -|v_S|$ (negative since the source [tuning fork] moves *away* from the observer), and $v_O = 0$ (since the observer is stationary), the Doppler-shifted frequency $f_O = 485$ Hz detected by the observer from the source of frequency $f_S = 512$ Hz is given by the relation $f_O = f_S(v + v_O)/(v - v_S)$ as

$$\frac{f_O}{f_S} = \frac{485\ \text{Hz}}{512\ \text{Hz}} = \frac{343\ \text{m/s} + 0}{343\ \text{m/s} - (-|v_s|)}$$

yielding

$$|v_S| = 343\ \text{m/s}\left[\left(\frac{512\ \text{Hz}}{485\ \text{Hz}}\right) - 1\right] = 19.1\ \text{m/s}$$

Thus, the first step in determining where the tuning fork is located when the observer hears the 485 Hz sound is to find how far the tuning fork must fall (starting from rest, with downward acceleration $g = 9.80\ \text{m/s}^2$) before it reaches the required speed of $|v_S| = 19.1$ m/s. Taking downward as positive, and using $v_y^2 = v_{0y}^2 + 2a_y(\Delta y)$, this displacement is found to be

$$(\Delta y)_1 = \frac{|v_S|^2}{2g} = \frac{(19.1\ \text{m/s})^2}{2(9.80\ \text{m/s}^2)} = 18.6\ \text{m}$$

The time required for the sound emitted by the tuning fork at this location to travel back to the release point is

$$\Delta t = \frac{(\Delta y)_1}{v_{\text{sound}}} = \frac{18.6\ \text{m}}{343\ \text{m/s}} = 0.054\,2\ \text{s}$$

During this time interval, the tuning fork will drop an additional distance of

$$(\Delta y)_2 = |v_S|(\Delta t) + \frac{1}{2}g(\Delta t)^2$$

$$= (19.1\ \text{m/s})(0.054\,2\ \text{s}) + \frac{1}{2}(9.80\ \text{m/s}^2)(0.054\,2\ \text{s})^2 = 1.05\ \text{m}$$

Thus, at the instant the observer at the release point hears the 485 Hz sound, the tuning fork is located

$$(\Delta y)_{\text{total}} = (\Delta y)_1 + (\Delta y)_2 = 18.6\ \text{m} + 1.05\ \text{m} = 19.7\ \text{m} \text{ below that point.}$$ ◊

37. A pair of speakers separated by a distance $d = 0.700$ m are driven by the same oscillator at a frequency of 686 Hz. An observer originally positioned at one of the speakers begins to walk along a line perpendicular to the line joining the speakers as in Figure P14.37. (a) How far must the observer walk before reaching a relative maximum in intensity? (b) How far will the observer be from the speaker when the first relative minimum is detected in the intensity?

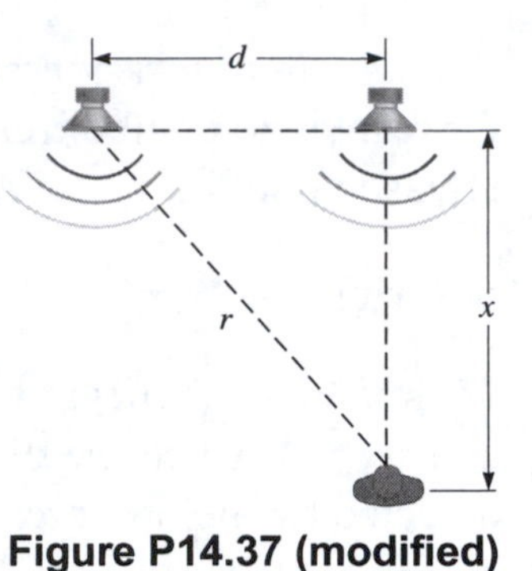

Figure P14.37 (modified)

Solution

Assuming the velocity of sound in the air is $v = 343$ m/s, the wavelength of the sound emitted by the speakers is

$$\lambda = \frac{v}{f} = \frac{343 \text{ m/s}}{686 \text{ Hz}} = 0.500 \text{ m}$$

(a) Relative maxima (constructive interference) in the combined sounds from the two speakers will occur when the difference in the path lengths (r and x in the figure given above) is an integral number of wavelengths. That is, when

$$\Delta d = r - x = n\lambda \quad \text{for } n = 0,\ 1,\ 2,\ 3,\ \ldots$$

In this case, the observer is always closer to the rightmost speaker, so the $n = 0$ case with $r = x$ cannot be achieved. Therefore, the first relative maximum detected will occur when $n = 1$ and $r = x + \lambda$. From the Pythagorean theorem, $r^2 = x^2 + d^2$, so we have $(x + \lambda)^2 = x^2 + d^2$. Expanding and simplifying this expression gives $2\lambda x = d^2 - \lambda^2$, and

$$x = \frac{d^2 - \lambda^2}{2\lambda} = \frac{(0.700 \text{ m})^2 - (0.500 \text{ m})^2}{2(0.500 \text{ m})} = 0.240 \text{ m} \qquad \Diamond$$

(b) For a relative minimum, the path difference must be an odd number of half-wavelengths, or $\Delta d = r - x = (2n + 1)(\lambda/2)$ for $n = 0,\ 1,\ 2,\ 3,\ \ldots$. Thus, at the first minimum, $n = 0$ so $r = x + \lambda/2$, and the Pythagorean theorem gives $(x + \lambda/2)^2 = x^2 + d^2$, which simplifies to $\lambda x = d^2 - (\lambda/2)^2$ and yields

$$x = \frac{(0.700 \text{ m})^2 - (0.250 \text{ m})^2}{0.500 \text{ m}} = 0.855 \text{ m} \qquad \Diamond$$

41. A stretched string of length L is observed to vibrate in five equal segments when driven by a 630-Hz oscillator. What oscillator frequency will set up a standing wave so that the string vibrates in three segments?

Solution

A standing wave pattern in a string fixed at each end consists of a series of uniformly spaced, stationary nodes (points of zero vibration) with an antinode (point of maximum vibration) halfway between each pair of adjacent nodes. The vibrating portions of string between successive nodes (called loops) have a uniform length of $\lambda/2$, where λ is the wavelength of the transverse waves traveling in the string.

Because a standing wave pattern must contain an integral number of loops, the string can produce a standing wave pattern (or resonance) only when its length is

$$L = n\left(\frac{\lambda}{2}\right) \quad \text{where } n = 1, 2, 3, \ldots$$

The wavelength, λ, and the frequency of the oscillations, f, are related to the speed of the waves by the expression $v = \lambda f$. The speed of transverse waves in a string is $v = \sqrt{F/\mu}$, and is constant as long as the tension F and the linear density μ do not change. Hence, in a given string, the frequencies that will produce resonance (and standing wave patterns) are

$$f_n = \frac{v}{\lambda_n} = \frac{v}{(2L/n)} = n\left(\frac{v}{2L}\right)$$

The ratio of the frequency that will form a standing wave of three loops ($n = 3$) to that which forms a standing wave of five loops ($n = 5$) is then

$$\frac{f_3}{f_5} = \frac{3\,\cancel{(v/2L)}}{5\,\cancel{(v/2L)}} = \frac{3}{5}$$

or

$$f_3 = \left(\frac{3}{5}\right)f_5$$

Thus, if for the given string, $f_5 = 630$ Hz, the oscillator frequency needed to form a standing wave of three loops is

$$f_3 = \frac{3(630\text{ Hz})}{5} = 378\text{ Hz}$$

◊

49. The windpipe of a typical whooping crane is about 5.0 ft. long. What is the lowest resonant frequency of this pipe, assuming it is closed at one end? Assume a temperature of 37°C.

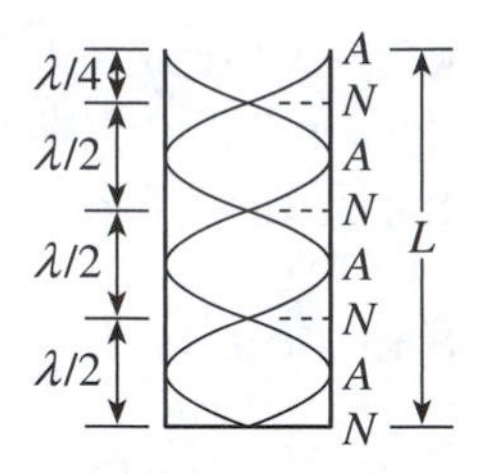

Solution

If we consider the windpipe of the whooping crane to contain an air column that is closed at one end, but open at the other end, it will be able to produce resonance for any frequency (and associated wavelength) that can form a standing wave pattern in the column with a node at the closed end and antinode at the open end. The sketch above represents a standing wave in the air column of a pipe open at the upper end and closed at the lower end.

The position of the antinodes are labeled A and the positions of the nodes are labeled N. Observe that it is possible to have the open end at the first antinode and the bottom of the pipe at the position of a node only if the length of the pipe is $L = \lambda/4$, or $L = 3\lambda/4$, or $L = 5\lambda/4$, etc. In general, this type of pipe can produce resonance only if

$$L = \frac{(2n+1)\lambda}{4} \qquad n = 0, 1, 2, 3, \ldots$$

Therefore, for a pipe of fixed length L, the wavelengths for which resonance can be produced are $\lambda_n = 4L/(2n+1)$, and the associated frequencies are

$$f_n = \frac{v}{\lambda_n} = (2n+1)\left(\frac{v}{4L}\right) \qquad n = 0, 1, 2, 3, \ldots$$

where v is the speed of sound in the air inside the pipe. At $T = 37°\text{C} = 310$ K, the speed of sound is

$$v = (331 \text{ m/s})\sqrt{\frac{T_K}{273}} = (331 \text{ m/s})\sqrt{\frac{310}{273}} = 353 \text{ m/s}$$

so the lowest resonant frequency ($n = 0$) of the windpipe of the whooping crane (with length $L = 5.0$ ft) is

$$f_0 = [2(0)+1]\left(\frac{353 \text{ m/s}}{4(5.0 \text{ ft})}\right)\left(\frac{3.281 \text{ ft}}{1 \text{ m}}\right) = 58 \text{ Hz}$$ ◊

55. In certain ranges of a piano keyboard, more than one string is tuned to the same note to provide extra loudness. For example, the note at 1.10×10^2 Hz has two strings at this frequency. If one string slips from its normal tension of 6.00×10^2 N to 5.40×10^2 N, what beat frequency is heard when the hammer strikes the two strings simultaneously?

Solution

In the fundamental mode of vibration, a string fixed at both ends has a node at each end and a single antinode at its center. Thus, the length of the string is $L = \lambda_1/2$ and the fundamental frequency of the string is

$$f_1 = \frac{v}{\lambda_1} = \frac{v}{2L}$$

where v is the speed of transverse waves in the string. This speed is given by $v = \sqrt{F/\mu}$, where F is the tension in the string and μ is the linear density of the string. If two strings have the same length and the same linear density, but different tensions F and F', the ratio of their fundamental frequencies is

$$\frac{f_1'}{f_1} = \frac{(1/2L)\sqrt{F'/\mu}}{(1/2L)\sqrt{F/\mu}} = \sqrt{\frac{F'}{F}}$$

or

$$f_1' = f_1\sqrt{\frac{F'}{F}}$$

When these two strings are sounded simultaneously, a beat frequency equal to the difference of their fundamental frequencies will be heard. For the two strings described in the problem statement, this beat frequency is

$$f_{\text{beat}} = f_1 - f_1' = f_1\left(1 - \sqrt{\frac{F'}{F}}\right) = \left(1.10 \times 10^2 \text{ Hz}\right)\left(1 - \sqrt{\frac{5.40 \times 10^2 \text{ N}}{6.00 \times 10^2 \text{ N}}}\right) = 5.64 \text{ Hz}$$ ◊

59. A student holds a tuning fork oscillating at 256 Hz. He walks toward a wall at a constant speed of 1.33 m/s. (a) What beat frequency does he observe between the tuning fork and its echo? (b) How fast must he walk away from the wall to observe a beat frequency of 5.00 Hz?

Solution

Initially, the student serves as a moving source $(v_S' = v_{\text{student}})$ of sound having $f_S = 256$ Hz. The wall acts as a stationary $(v_O' = 0)$ receiver or observer of this sound. The sound wave received and reflected by the wall has frequency

$$f_{\text{echo}} = f_O' = f_S\left(\frac{v + v_O'}{v - v_S'}\right) = f_S\left(\frac{v}{v - v_{\text{student}}}\right)$$

where $v = 343$ m/s is the speed of sound in air.

At the reflection, the wall serves as a stationary source $(v_S = 0)$ of sound having frequency f_{echo}, and the student is a moving observer $(v_O = v_{\text{student}})$ of this sound. The frequency of the echo heard by the student is

$$f_O = f_{\text{echo}}\left(\frac{v + v_O}{v - v_S}\right) = \left[f_S\left(\frac{v}{v - v_{\text{student}}}\right)\right]\left(\frac{v + v_{\text{student}}}{v}\right) = f_S\left(\frac{v + v_{\text{student}}}{v - v_{\text{student}}}\right)$$

The beat frequency heard by the student is then

$$f_{\text{beat}} = |f_O - f_S| = \left|f_S\left(\frac{v + v_{\text{student}}}{v - v_{\text{student}}}\right) - f_S\right| = 2f_S\left|\frac{v_{\text{student}}}{v - v_{\text{student}}}\right|$$

(a) If the student *approaches* the wall at speed $|v_{\text{student}}| = 1.33$ m/s, the needed velocity is $v_{\text{student}} = +1.33$ m/s, and the beat frequency is

$$f_{\text{beat}} = 2f_S\left|\frac{v_{\text{student}}}{v - v_{\text{student}}}\right| = 2(256\text{ Hz})\left|\frac{+1.33\text{ m/s}}{343\text{ m/s} - 1.33\text{ m/s}}\right| = 1.99\text{ Hz} \quad \lozenge$$

(b) If the student *moves away* from the wall at speed $|v_{\text{student}}|$, the needed velocity is $v_{\text{student}} = -|v_{\text{student}}|$. If the student is to hear $f_{\text{beat}} = 5.00$ Hz, then

$$2f_S\left|\frac{-|v_{\text{student}}|}{v + |v_{\text{student}}|}\right| = 5.00\text{ Hz} \qquad \text{or} \qquad \frac{2f_S|v_{\text{student}}|}{v + |v_{\text{student}}|} = 5.00\text{ Hz}$$

This yields

$$|v_{\text{student}}| = v\left(\frac{5.00\text{ Hz}}{2f_S - 5.00\text{ Hz}}\right) = \left(343\ \frac{\text{m}}{\text{s}}\right)\left[\frac{5.00\text{ Hz}}{2(256\text{ Hz}) - 5.00\text{ Hz}}\right] = 3.38\text{ m/s} \quad \lozenge$$

66. A student uses an audio oscillator of adjustable frequency to measure the depth of a water well. He reports hearing two successive resonances at 52.0 Hz and 60.0 Hz. How deep is the well?

Solution

The well will act as a pipe closed at the lower end and open at the upper end. At resonance, the sound waves form a standing wave pattern with a node at the bottom of the well and an antinode at the top of the well. Since the distance from a node to the nearest antinode is a quarter wavelength, the longest resonant wavelength (i.e., the fundamental) which can meet the requirements of a node at the bottom and an antinode at the top is $\lambda_1 = 4L$, where L is the depth of the well. The lowest resonant frequency of the well is then $f_1 = v/\lambda_1 = v/4L$, where v is the speed of sound in air.

In a standing wave pattern, the distance between adjacent nodes is a half-wavelength. Thus, when the well is resonating in a higher harmonic, the distance from the node at the bottom of the well and the uppermost node must be an integer multiple of $\lambda/2$. The distance from the uppermost node to the antinode at the top of the well is $\lambda/4$, so the total length of the well is

$$L = n\frac{\lambda}{2} + \frac{\lambda}{4} = (2n+1)\frac{\lambda}{4} \qquad (n = 0,\ 1,\ 2,\ 3,\ \ldots)$$

or equivalently

$$L = (2n-1)\frac{\lambda}{4} \qquad (n = 1,\ 2,\ 3,\ \ldots)$$

The resonant wavelengths of this pipe, closed at one end and open at the other, are

$$\lambda_n = \frac{4L}{(2n-1)} \qquad n = 1,\ 2,\ 3,\ \ldots$$

and the resonant frequencies are

$$f_n = \frac{v}{\lambda_n} = (2n-1)\frac{v}{4L} = (2n-1)\,f_1 \qquad n = 1,\ 2,\ 3,\ \ldots$$

Observe that this says the possible resonant frequencies of the well are f_1, $3f_1$, $5f_1$, … That is, only the *odd integer* multiples of the fundamental frequency are resonant frequencies for a pipe closed at one end and open at the other, and the difference between any two successive resonant frequencies is $2f_1$. Thus, if the well is observed to have successive resonant frequencies of 52.0 Hz and 60.0 Hz, we know that

$$2f_1 = 2\left(\frac{v}{4L}\right) = 60.0\ \text{Hz} - 52.0\ \text{Hz} = 8.00\ \text{Hz}$$

and the depth of the well is

$$L = 2\left[\frac{v}{4(8.00\ \text{Hz})}\right] = \frac{343\ \text{m/s}}{16.0\ \text{Hz}} = 21.4\ \text{m}$$

◊

74. A student stands several meters in front of a smooth reflecting wall, holding a board on which a wire is fixed at each end. The wire, vibrating in its third harmonic, is 75.0 cm long, has a mass of 2.25 g, and is under a tension of 400 N. A second student, moving towards the wall, hears 8.30 beats per second. What is the speed of the student approaching the wall?

Solution

When the wire, fixed at both ends and of length $L = 0.750$ m, vibrates in its third harmonic, the wavelength of the standing wave in the wire is

$$\lambda = \frac{2L}{3} = \frac{2(0.750\text{ m})}{3} = 0.500\text{ m}$$

The speed of transverse waves in the wire is given by

$$v_{\text{wire}} = \sqrt{\frac{F}{\mu}} = \sqrt{\frac{F}{m/L}} = \sqrt{\frac{FL}{m}} = \sqrt{\frac{(400\text{ N})(0.750\text{ m})}{2.25\times10^{-3}\text{ kg}}} = 365\text{ m/s}$$

The frequency at which the wire vibrates, and hence the frequency of the sound it generates in the surrounding air, is

$$f_S = \frac{v_{\text{wire}}}{\lambda} = \frac{365\text{ m/s}}{0.500\text{ m}} = 730\text{ Hz}$$

Since both the wire and the wall are stationary, the frequency of the reflected wave from the wall is the same as the frequency of the wave coming directly from the wire. The student will then hear sound from two stationary sources $(v_S = 0)$, both emitting a frequency $f_S = 730$ Hz. Walking at speed $|v_{\text{student}}|$, the student walks *away* from the wire (source 1) with velocity $v_{O,1} = -|v_{\text{student}}|$, and *toward* the reflecting wall (source 2) with velocity $v_{O,2} = +|v_{\text{student}}|$. Thus, the sound coming directly from the wire to the student is Doppler shifted to a lower frequency while the sound reflecting from the wall is Doppler shifted to a higher frequency. The beat frequency the student will detect while hearing these two sounds simultaneously is

$$f_{\text{beat}} = f_{O,2} - f_{O,1} = f_S\left(\frac{v_{\text{air}} + v_{O,2}}{v_{\text{air}} - v_{S,2}}\right) - f_S\left(\frac{v_{\text{air}} + v_{O,1}}{v_{\text{air}} - v_{S,1}}\right) = f_S\left(\frac{v_{\text{air}} + |v_{\text{student}}|}{v_{\text{air}} - 0}\right) - f_S\left(\frac{v_{\text{air}} - |v_{\text{student}}|}{v_{\text{air}} - 0}\right)$$

or $f_{\text{beat}} = 2f_S|v_{\text{student}}|/v_{\text{air}}$. If it is observed that $f_{\text{beat}} = 8.30$ Hz, the speed of the walking student is

$$|v_{\text{student}}| = \frac{v_{\text{air}} f_{\text{beat}}}{2f_S} = \frac{(343\text{ m/s})(8.30\text{ Hz})}{2(730\text{ Hz})} = 1.95\text{ m/s}$$

◊